FIRE EATERS

FIRE EATERS

THE PEOPLE AND AIRCRAFT COMBATTING WILDFIRES OVER THE LAST CENTURY

PETER PIGOTT

Publisher: Meghan Macdonald | Acquiring editor: Kathryn Lane | Editor: Dominic Farrell
Cover designer: Laura Boyle
Cover image: planes: Pierrick Caillet; forest fire: Unslash/Matt Howard

Library and Archives Canada Cataloguing in Publication

Title: Fire eaters : the people and aircraft combatting wildfires over the last century / Peter Pigott.
Names: Pigott, Peter, author
Description: Includes bibliographical references and index.
Identifiers: Canadiana (print) 20240537319 | Canadiana (ebook) 20240537335 | ISBN 9781459755079 (softcover) | ISBN 9781459755093 (EPUB) | ISBN 9781459755086 (PDF)
Subjects: LCSH: Ontario. Provincial Air Service—Equipment—History. | LCSH: Airtankers (Forest fire control)—Ontario—History. | LCSH: Aeronautics in wildfire control—Ontario—History. | LCSH: Air pilots—Ontario—History. | LCSH: Wildfires—Ontario—History.
Classification: LCC SD421.43 .P54 2025 | DDC 634.9/61809713—dc23

We acknowledge the support of the Canada Council for the Arts and the Ontario Arts Council for our publishing program. We also acknowledge the financial support of the Government of Ontario, through the Ontario Book Publishing Tax Credit and Ontario Creates, and the Government of Canada.

Printed and bound in Canada.

Dundurn Press
1382 Queen Street East
Toronto, Ontario, Canada M4L 1C9
dundurn.com, @dundurnpress

For Aria Adamson, Ethan and Cole White, Jack and Josie Rixon, and Tommy Lennox, who was born on February 28, 2024. Do better than we did.

WE DID OVER TEN HOURS IN THE AIR TODAY, A BATTLE FOR EVERY MINUTE. Entire olive groves turned to ash in minutes; in the time it took to refuel, a whole mountain burned. I've never seen such ferocity — the fire just vaporizing everything in its path.

Yesterday, I was dropping water on village houses that were burning, totally heartbreaking. It was olive groves, honeybee hives. I could see about thirty beehives go up like torches. Full of wax and sugar, they burn like firecrackers. People who had very little to begin with now losing everything.

So today, when the fire was running straight toward a village, I was asked to try to hit its head near a dry riverbed. Initially, we try to contain the fire by attacking the head — this is the part moving with the direction of the wind. Its mouth is where the oxygen and fuel are being devoured. Truthfully, it didn't look too bad. As lead plane, I lined up for a drop on the left flank of the fire's head that was climbing the hill and about to run down. I entered light smoke, planned to roll my load across the ridge, and pitched down along the flank.

Then I received a very unwelcome surprise. As I came in to drop parallel to the fire head, a flock of massive storks and pelicans flew up directly into my flight path! They had been feasting on the insects and small mammals fleeing the fire, an easy meal for them but nearly a disaster for me. Just one of the ever-present and unexpected dangers when flying close to the ground.

As soon as I hit the drop button, I cleared the smoke, which was from the faster-burning grass and shrubs. At that moment, I encountered the huge vortex of the head devouring the heavy fuels on the other side of the hill. I knew immediately I was in trouble....

We had our asses handed to us by Mother Nature that day. I've never felt so defeated and humiliated in my life. But I'll try again tomorrow!

— Brad Blois, Canadian airtanker pilot in Rhodes, Greece

Contents

Preface

I WAS ONCE ON JOURNALISTIC ASSIGNMENT IN CROATIA. TAKING THE FERRY from Zadar, I joined everyone on deck, all of us watching a forest fire on one of the islands. Someone shouted "Kanadini!" And on cue, like the cavalry, three yellow water bombers swooped down on the fire, dropped their loads, and climbed away.

Never losing an opportunity to show off my aviation knowledge, especially to all the Italian girls, I remarked, "Those are our Canadian CL-215s."

The old lady beside me gestured and ranted something. One of the girls laughed and translated, "She says no, those are our angels."

Although I was unaware of it then, *Fire Eaters** began that day.

• • •

In the autumn of 2023, Canadians were tentatively beginning to grasp the magnitude of what had taken place in their country that summer. The estimate of the number of hectares burned by fire exceeded seventeen million — a number that multiplied as more accurate mapping became available. On

* Unless otherwise noted, all interviews in the text were conducted by the author.

Three CL-415s execute a scoop with precision and control in the choreography of water bombing.

May 6, Alberta had declared a Provincial State of Emergency due to wildland fires, followed by Nova Scotia on June 1. On June 7, New York City had the worst air quality in the world. The sky turned auburn as smoke from wildfires in Canada spread throughout the boroughs. By June 21, the Donnie Creek Wildfire in British Columbia had broken the record for the single largest fire in the province.

The United States and Canada signed an agreement on June 23 that allowed Canada access to sophisticated U.S. satellite data to identify wildfires. On June 27, about midway through the normal fire season, the record for area burned nationally, forty-two million acres (seventeen million hectares), set in 1989, was broken and would continue to grow. Fort Good Hope, just

145 kilometres south of the Arctic Circle, experienced a scorching 37.4° C (99.3° F) day on July 8, marking the northernmost occurrence of such a temperature ever recorded. The Northwest Territories (NWT) declared a Territorial State of Emergency on August 14. Approximately 70 percent of the territory, including its capital, Yellowknife, was evacuated to neighbouring provinces. On August 18, British Columbia declared a Provincial State of Emergency as fire activity increased across the province.

For perspective, Canada's burned area that summer surpassed the combined area of Switzerland and Greece and was the size of Uruguay or Florida. In Quebec alone, wildfires had burned an area larger than Switzerland, and thirty thousand people, mostly Indigenous, were evacuated.

In that summer of 2023, in the vast boreal forests of Canada, considered one of the world's biggest terrestrial carbon vaults, more than 1.7 billion tons of planet-heating gases were released by the wildfires — about three times the total emissions of this major fossil fuel–producing nation each year. And it wasn't only in Canada. That summer, a firestorm in Lahaina, Hawaii, killed at least one hundred people. Greece suffered the largest single wildfire in European Union (EU) history, the wildfires producing some twenty megatons of CO2 emissions. I tried not to think too hard about what that meant for the future of the youngsters to whom this book is dedicated.

By the mid-2030s, the planet will be two degrees hotter. For Canada, that means four degrees hotter, as the blanket of snow and ice that has reflected the sun's heat back out of the atmosphere will no longer be there. Our northern forests absorbing more heat means wildfires will burn hotter, larger, and longer, poisoning the air for urban millions. In an apocalyptic scenario, doomsayers foretell crop failures, drought, floods, mass migration to more moderate climes, and seashores receding, leading to a decline in prosperity, social cohesion, and physical health.

The 49th parallel serves as the southern border for Canada's western provinces, but more than half of the Canadian population lives south of it in eastern Canada. The vast northern forests and tundra, out of sight and unmapped until the 1930s, were where forest fires historically occurred. Too far away and complex for those living in the southern part of the country to conceptualize, and few apart from the Indigenous Peoples living in them, they were little regarded. Conflagrations happened to other people far, far away — not in cities like Fort McMurray and Jasper, Alberta; West Kelowna, British Columbia; or suburban Halifax.

While there is a general increase in the number of fires as a result of climate change, such fires have been localized. However, the smoke produced by them spreads far beyond where the burns occur. Smoke doesn't care about borders. Even in the short term, exposure to wildfire smoke can damage the lungs and heart, cause strokes, and exacerbate asthma and other respiratory issues. Our bodies evolved to thrive within a stable temperature range. "Because living things share one simple fate: if the temperature they are used to — what scientists call their 'Goldilocks Zone' — rises too high too fast, they die."[1]

To avoid this, humans flee or migrate. And it is not only humans who will be migrating en masse to find cooler climes. The world's wildlife also faces profound, catastrophic changes. The boreal forests are a trove of habitats for creatures such as moose, bears, coyotes, and songbirds. Rising temperatures affect vegetation, food sources, access to water, and much more. Increases in temperature have already triggered the collapse of fragile ecosystems, like coral in the Great Barrier Reef. Ecosystems will become uninhabitable for certain animals, especially those that cannot sweat, forcing wildlife to migrate outside of their usual patterns in search of food and liveable conditions.

The following year wasn't much better. By July 31, 2024, in the United States alone, 28,154 wildfires had burned almost two million acres (eight hundred thousand hectares). More than twenty-eight thousand wildland firefighters were fighting them, with 1,781 engines and numerous aviation resources, including 164 helicopters. There was extreme fire behaviour across multiple geographic areas, with evacuation orders in effect on twenty-two wildfires. By July 24, the Park Fire had burnt more than 395,000 acres (160,000 hectares) in northern California to become the fourth largest in that state's history. Both it and the Jasper fire in Canada, which had destroyed one-third of the resort town, had spawned ominous "pyrocumulonimbus" clouds, towering formations that spat lightning, potentially starting more fires. And it was only the middle of the fire season ...[2]

Stephen Pyne, a fire historian at Arizona State University, warned, "We are watching mythology become ecology — it's a slow-motion Ragnarök. We've had ice ages in the past, but we are now living through what I call the 'Pyrocene.' Imagine an ice age, but instead of ice as a forming feature, we have fire."[3]

Author John Vaillant has christened this era "Petrocene Age," describing it as one characterized by our addiction to fossil fuels, particularly oil. With the dawn of the Industrial Revolution, humans began burning more and more ancient biomass in the form of coal, oil, and natural gas to power everything from locomotives and ships to factories and cities, and their combustion altered the climate. These actions, along with changes in land use, led to a rise in greenhouse gases, which trapped heat around the planet. Combined with widespread deforestation by settlers and lumber companies, they had pumped heat-trapping emissions — carbon dioxide, methane, nitrous oxide, and hydrofluorocarbons — into the atmosphere. By the mid-twentieth century, the bill for the Industrial Revolution, with its reliance on fossil fuel, came due. The world is now grappling with the effects of climate change, including an alarming increase in the number and size of wildfires.

Our species invented the spear, the wheel, the printing press, eradicated smallpox, split the atom, landed on the Moon and Mars, and invented handheld devices that allow us to communicate worldwide. Our ingenuity and determination showed that we could do anything. How then could we not prevent wildfires, stop the glaciers from melting, the seashore from receding, or cleanse our cities of toxic air? More and more, fire is shaping the planet as cause, consequence, and catalyst. Unable to live without the energy that ancient fires in fossil fuels gave us, we alienated ourselves from Nature, and in the end, Nature outran us. The world we thought we understood changed around us because we changed it.

Al Gore, writing in the *New York Times* in 2019, put it starkly: "This is our generation's life-or-death challenge. It is Thermopylae, Agincourt, Trafalgar, Lexington and Concord, Dunkirk, Pearl Harbor, the Battle of the Bulge, Midway, and Sept. 11."[4]

• • •

I am an aviation author, not an environmental one, and at its heart, *Fire Eaters* is about the people and aircraft that have contained wildfires. Much of the book takes place in the twentieth century, the heyday of aggressive fire suppression, in all its perversity. Those I interviewed, airtanker and scooper aircrew, aircraft maintenance engineers, and smokejumpers alike, possess skills and resilience that I can't even fathom.[5]

"I was back on the ground," Cliff Lynn of Neptune Air Services said, "looking at the smoke on the hill, and you see all these powerful men just sitting there in silence, thinking about what they had just experienced. They were protecting schools, homes, and other buildings — and they had to do it fast. This is the kind of fire that people remember because you see how powerful fire is, and you realize that there are times when you are completely helpless."

Ray Horton, a long-time airtanker pilot, asked me to make clear to readers that contrary to popular belief, airtankers and water scoopers *do not* put out fires. That award, he said, goes to the hardworking ground crews that ultimately must ensure that a fire is contained and extinguished, all the while faced with difficult terrain fraught with hazards such as rocks and falling trees. Fire Boss pilot Brad Blois added, "All we do is slow things down, cool things off, or prevent jumps in the fire line so that the heroes on the ground can do their jobs."

By the 2040s, when those this book is dedicated to will be entering adulthood, the climate change bailiffs will be at the door. The aviator/author Antoine de Saint Exupéry reminded us that we haven't inherited the Earth from our parents, but rather, we are borrowing it from our children. By accident of fate, we are now at a critical moment in the history of our planet. There is a lot of fire coming at us. We can fight it and lose. Or we can turn what has become a relentless enemy back into a necessary friend. Our choice — our children's future …

CHAPTER 1

"Some Say the World Will End in Fire"[1]

THE HISTORIC CONFLAGRATIONS OF THE PAST — THE 1835 GREAT FIRE OF New York, the 1871 Great Fire of Chicago, the 1886 Vancouver Fire, the 1900 Great Fire that swept through Canada's capital, and the Toronto Fire of 1904 were urban tragedies. In each city, those in power responded by putting in place measures to ensure that such fires would never occur again. Professional firefighting companies were created to suppress those blazes that did break out. Policies were put in place to minimize the likelihood of other fires spreading, high-pressure water mains to supply fire hydrants were installed, and building codes that insisted on the replacement of wood with fire-resistant brick and structural steel were put in place. Whole cities would never be consumed by fire again. As Christopher Wren rebuilt much of London after its Great Fire in 1666, from the ashes of those urban fires arose the phoenix of the modern city.

But until the invention of aircraft, when wildfire threatened, our only defence was to run and to pray for rain — in that order. Until the use of aircraft to detect and suppress wildfire, this had not changed in millions of years. However mighty or sophisticated the civilization, the wall of flames coming toward it was unstoppable. There was only one option: flight — not fight.

Until a hundred years ago, there was nothing humanity could do about wildfires — like epidemics (before municipal sanitation and antibiotics), hurricanes, and earthquakes — they were accepted hazards and out of human control.

What altered urban and political attitudes to wildfires in North America were three record conflagrations, two in the United States and one in Canada. The "Yacolt Burn of 1902," named after the community at its western edge, was the largest wildfire recorded in Washington's history. "Imagine flames three hundred feet tall, wind whipping at forty miles per hour, and sparks leaping over 1/2 mile, spreading fire twenty miles in twelve hours. At the end of three days, 238,000 acres of forested land lay smoldering in ash and smoke."[2] Thirty-eight people died, and 148 families lost their homes.

In Idaho and Montana, the year 1910 was the driest on record. Ignited perhaps by the cinder-spewing locomotives of the Chicago, Milwaukee, and Puget Sound Railway or set intentionally by agents of the unscrupulous William A. Clark, "The Copper King," a wildfire, raged for two whole days and nights, August 20 and 21. It had crossed 1.2 million hectares of virgin timberland in northern Idaho and western Montana by noon on the twenty-first. Daylight was said to be dark as far north as Saskatoon, as far south as Denver, and as far east as New York. It was like nothing anyone had ever seen, and the sensationalist newspapers of the day reported that many thought the end of the world was nigh.

Commemorated as the "Big Burn," the 1910 fire claimed eighty-six victims. The Miramichi Fire in 1825 had spread over much of New Brunswick and killed unknown hundreds; in the Peshtigo, Wisconsin, fire in 1871, the estimates placed the deaths at more than two thousand people; the Thumb Fire, Michigan, in 1881 claimed 282 victims; and in 1894, in the Hinckley, Minnesota fire, 418 people died.

None of those conflagrations, awful as they were, made enough of an impression to spur governments to consider ways to tackle them. However, the "Big Burn" of 1910 occurred at an opportune time in history, and it scorched its way into the North American conscience as no other wildfire had done before.

It was the spread of information technology that allowed this to occur; the railway networks, the telegraph, and telephone all ensured that this fire entered the public consciousness. The internet of the day, newspapers were at their height of influence, and never was a distant forest fire given so much exposure — and in headlines so large. Not only was this the first wildfire that the five-year-old United States Forest Service (USFS) would fight, but it happened in a decade of unprecedented optimism and progress: there were now Ford Model Ts, dirigibles, telephones, elevators, silent movies, burgeoning cities with electric lights. In Belfast, Northern Ireland, the largest ship ever built was about to be launched, the Panama Canal had been completed, and one Louis Bleriot had flown across the English Channel! There was nothing, it seemed, that could not be solved by science and the internal combustion engine.

Most of those who died in "Big Burn" were the first rangers of the Forest Service. In the railway, mining, and brothel-ridden towns, rangers were the most hated men of what was the still-Wild West. Unpopular with loggers, hunters, saloon keepers, and settlers, they were poorly paid and had to buy their own equipment. A hunter shot a ranger, claiming he'd mistaken him for a deer. A mob, enraged by new limits on plundering the forests, hanged Gifford Pinchot, the first Forest Service chief, in effigy. In Congress, foes of the agency lobbied for felling entire forests before they caught fire. "Not one cent for scenery!" cried the speaker of the House, Joseph Cannon, one of many legislators opposed to President Theodore Roosevelt's conservationist agenda. "Trees, after all," he proclaimed, "are just boards and railroad ties in waiting."

President Roosevelt and Pinchot worked to save forests from the axe and fire, and the "Big Burn" gave them the platform they needed to convince the government to allocate more funds to train wildfire fighters; until then, they had been largely recruited from saloons or were drifters. The "Big Burn" also gave the public an authentic hero in the form of Edward Crockett Pulaski. When wildfire blocked his men's escape, he led them and their horses to the Nicholson mine. He ordered the crew into the mineshaft and fought the raging fire at the entrance with his bare hands, holding his panicked men off at the point of a gun. Big Ed not only gave real-life authenticity

Men fighting wildfires, 1916. If wildfires were fought at all, it was by "pick-up crews," so called because they were drifters picked up off the streets or press-ganged.

to containing wildfires but also publicized what the new guardians of the national forests did — until then, their work was unknown to the world.

Human habitations that encroached on what were thought of as unlimited forests — unlimited, like the many species of animals that were hunted to extinction or near-extinction — needed to be protected from fire. Where the Europeans saw the North American wilderness as something to own and domesticate, Indigenous Peoples, who lived in harmony with Nature, understood that wildfires were part of the natural ecosystem and that forests and grasslands needed frequent low-intensity burning to remove accumulated amounts of dead brush.

With colonization, *cikilaxwm*, the Indigenous strategy for forest stewardship, which had been practised for thousands of years, was criminalized, to be replaced by strategies endorsed by governments, the railways, and logging companies. That colonization of North America was accompanied by genocide and the theft of most of the land from Indigenous Peoples is well known. But as M.R. O'Connor writes, colonization disrupted the traditional land management practices that Indigenous Peoples practised, including fire-lighting. This had a profound effect on the forests of North America. "As early as 1634, European settlers on the continent were noticing that the death of Native Americans from disease was changing the now unattended landscape." In the ultimate paradox, the practice of suppressing wildfires allowed a highly combustible bank of fuel to build up, which served to fuel the megafires of the present.[3]

The United States developed a strong culture of fire suppression, which would define its strategies throughout the twentieth century. Wildfire was anthropomorphized, i.e., given human attributes, and had to be "killed" by choking, drowning, and suffocating. The policy was that you don't negotiate with the enemy, whether Nazis, terrorists, or wildfires; you bomb (suppress) them into submission. It made fires easier to comprehend, and fighting them was reimagined as a campaign, demanding militaristic solutions, and originally, it was the Buffalo Soldiers that suppressed them.[4] The military didn't know about fighting fires, but they could follow orders. During the First World War, when leaders feared enemy saboteurs could start hundreds of massive wildland fires, overwhelming the country's firefighting capabilities, forest fire management took on a national security perspective.

So too did the deadlines imposed on those tasked with extinguishing forest fires. To prevent future "Big Burns," in 1935, the chief of the Forest Service, Ferdinand A. Silcox, announced the "10:00 a.m. policy." All fires on Forest Service land were to be completely extinguished by the morning after the day they were reported. It was a well-meant but flawed policy, and he has been vilified for it. But it was during the Depression, and through the Civilian Conservation Corps (CCC) and Works Projects Administration (WPA) projects in the national forests, he provided employment for thousands of hungry men. No one understood that fire suppression was wrecking the forests it was intended to save. Though many blazes pushed past the deadline, the 10:00 a.m. mandate remained in effect until the late 1970s. Fortunately, the pendulum has begun to swing back, and forestry management practices now acknowledge the benefits of fire in the forest ecosystem. Today, the goal is to fight fire with fire — to manage it rather than rush to extinguish it, and as the Indigenous Peoples had always done, cultivate more prescribed burning.

In 1942, a Japanese submarine shelled the Ellwood Oil Field on the California coast near the Los Padres National Forest. The fear was that incendiary shells could touch off wildfires at a time when the war effort had depleted firefighter ranks. The Wartime Ad Council put out the slogans "Forest Fires Aid the Enemy" and "Our Carelessness, Their Secret Weapon." And from his movie *Bambi*, Walt Disney authorized the depiction of a helpless fawn in danger of incineration. The deer was replaced in 1945 by the most recognizable and endearing symbol the U.S. Forest Service has ever produced, Smokey Bear, with his message: "Only YOU Can Prevent Forest Fires."[5] Embedded in the minds of the Cold War generation of schoolchildren was that total fire suppression was the moral equivalent of fighting the nation's enemies bent on destroying America's way of life.

The horror of the Soviet Union dropping incendiary bombs (such as napalm or thermite) with greater firepower than the wartime Japanese balloon bombs and targeting the highly flammable forests made suppressing wildfires tantamount to civil defence. Advocates of the 1954 "Operation Firestop" conference took up the science and weapons that had won the Second World War to fight wildfires at home. The U.S. Forest Service and the Bureau of Land Management obtained mothballed military aircraft from the General Services Administration, converted them into airtankers, and sold them to firefighting companies, which they would then contract each summer. Although an aviation colossus, there was no incentive in the United States for its aircraft manufacturers to design and build aircraft solely to be water scoopers and airtankers.

In this era, with forest management practices that include the use of prescribed fires, blazes ignited using aerial ignition dispensers, and a more relaxed attitude to lightning fires, which are allowed to play their natural role (as long as they bring no harm to humans), Smokey's message now seems counterproductive and outdated. Not so, the Forest Service is quick to point out: "Smokey still wants you to put out fires, but he's talking about the wildfires that humans start accidentally or through irresponsible behavior."

• • •

That fires in the boreal region burned more frequently in the twentieth century was because forests were no longer solely the haunt of a few loggers and Indigenous people. In the prosperity of the era, with improved road and rail access, tourists, hikers, picnickers, and campers were lured to "discover" them. They expected — indeed, felt they were entitled to — neat picture-postcard vistas of closely packed conifers to romanticize over. To achieve these, forest management strategies ignored the ecological necessity of thinning forests, and soon, densely packed trees competed for resources. A century of making overstocked forests look like overgrown theme parks, followed by mountain pine beetle infestation, left behind too many aging trees and little space for the more resilient ones to be regenerated. The altered landscape now far more susceptible to high-intensity fires in a rapidly warming world, it was an ecological disaster in the making.

After the Second World War, shifting population patterns complicated the challenges faced by professional firefighters. Governments allowed — indeed, encouraged — us to settle in potentially dangerous places. Wildland fires did not encroach on communities. Rather, it was the communities, with government encouragement, that impinged on wildlands, where fires played their ecological role. The aesthetic appeal of residing and working amidst mountains and forests, in Paradise, California, and industrial towns like Fort McMurray, Alberta, attracted communities to build in areas prone to wildfire.

Wrongly identified as wildlands with houses in them, Paradise, Fort McMurray, Yellowknife, Lytton, Fort Good Hope, etc., are in reality urban enclaves within forests. The 2016 wildfire at Fort McMurray forced ninety thousand people to evacuate on the single road out of town. It destroyed more than two thousand homes and caused billions of dollars in damage. That the "Fort McMoney" evacuees drew their livelihood from the Alberta oil sands was a cruel irony. The city had been carved out of the boreal forest, and its location was key to the catastrophe. The wildfire had jumped from the encircling forests into the neighbourhoods of Centennial Trailer Park, Prairie Creek, and Gregoire, all in the Wildland Urban Interface (WUI). Paradise was doomed in 2018 because it had been built upon a system of volcanic ledges near a deep canyon in the Sierra Nevadas that bellowed out windstorms. The irony was that wildfire had not invaded them — they had invaded wildfire. The WUI is a beautiful place to live, until it goes feral.

Our present communities and transportation systems were designed for an environment that no longer exists. In the postwar expansion of exurbs into forests, no one had figured out that the active suppression of wildfires over the many decades would transform the undergrowth outside our backyards into incendiary bombs. In the summer of 2024, even after the city of Jasper had prepared itself with prescribed burns, the thinning of diseased forests, and the installation of sprinkler systems, a third of the town was reduced to ashes. As with the Park Fire, there was no way of combatting extended heat waves that left the forests tinder-dry. Compared with cool air, warm air holds more moisture — about 7 percent more per degree Celsius — and so draws more water from vegetation on the ground. Even a healthy forest without the infected trees would have been consumed by the Jasper fire, which roared into town with a wall of flame one hundred metres high. In July 2024, it was still flight, not fight …

• • •

President Teddy Roosevelt, the larger-than-life conservationist, provided guidance for Canada's own agenda in forest management. In 1906, Prime Minister Wilfrid Laurier formally announced the inauguration of government-sponsored forest conservation following the Canadian Forestry Convention.[6]

One of the world's most abundantly forested countries, Canada has 1.2 billion acres (453 million hectares) of forest, which account for almost 42 percent of its land base. Some of the earliest accounts of forest fires in what was to be Canada were written by Jesuit priests serving at various missions in New France. In 1735, Father Jean Pierre Aulneau wrote that the smoke from forest fires was so thick, it prevented him from "even once catching a glimpse of the sun" as he travelled from Lake Superior to Fort St. Charles, Minnesota, on Lake of the Woods. J.D. Cameron, the Hudson's Bay Company trader at Rainy Lake, Ontario, wrote that the years 1803 and 1804 were major fire years along the Minnesota-Ontario border country.

For the first fifty years following Confederation, forest fires were mostly left to burn unhindered. Unlike Washington, Ottawa did not control large tracts of forest. That meant it lacked a centralized federal force to fight wildfires. (It still lacks one.) With 97 percent of the forests owned either by the provinces or the two railways, the Canadian Pacific and Canadian National, there was little the federal government could do. Every province had a northern line of control beyond which they let fires burn. If wildfires were fought at all, the battles were waged by work parties marshalled by railway and timber mill managers who saw their profits going up in smoke — literally.

The success of those crews was severely limited by the difficulty they faced accessing the fires. Canoes were the standard mode of travel in northern Quebec and Ontario; in the more settled regions, however, patrol by foot, horseback, and railway was possible. In 1918, the St. Maurice Forest Protective Association drew media attention by introducing a motorcycle patrol.

"Pick-up crews," so called because they were made up of drifters, many of them recent immigrants, were picked up off the streets and press-ganged into service. These men were herded to the nearest railhead and then marched to the fire. Fire lines were hacked out with crosscut saws, long-handled shovels, the hazel hoe, and Pulaskis.[7]

Photographs from Library and Archives of Canada show the forest firefighters of the period in everyday cotton and wool clothing. One wonders how exhausted, untrained men were able to create (in modern jargon) "fuel breaks," which required digging down to the mineral soil around the fire. Hand pumps with water bags would not come into use in Canada until 1923. Fighting the monster must have been futile, but until the use of aircraft, it was all there was.

The poor access to remote areas meant that provincial governments were unaware of most wildfires, and even those it was made aware of were usually just left to burn, especially in northern regions inhabited only by the First Nations. To be fair, provincial governments were caught between encouraging colonization, i.e., the slash-burning of forest land by settlers seeking to farm it, and reassuring businesses, i.e., the forestry companies and railways, both of which regularly ignited the forests. Periodic conflagrations destroyed settlements, killed hundreds, and burned thousands of hectares of forest land across Canada. Never was this more evident than on the single day of May 19, 1870, when a wildfire, begun by settlers' slash-burning in the Quebec Saguenay-Lac-Saint-Jean region, swiftly consumed everything within 150 miles, leaving hundreds of families destitute.

If governments were unanimous in regarding forest fires as an enemy that had to be fought, their capacity to do so was limited. The fact that those most affected by the fires were Indigenous people, who were then not enfranchised, made it all too easy to restrain expenditures on forest protection. Some efforts were made. Fire towers were built, and rangers (*garde-feux* in Quebec) were hired to patrol and levy fines — but to little avail.

The smoky haze from rural fires that drifted each August over Canadian towns and cities was an accepted way of life. If anything, it symbolized the progress of a growing young country. Settlers generated slash (debris left from burns to clear land); the timber industry (as it shifted from sawtimber to pulp) allowed debris from its mills to smolder; and locomotives pushed their way through tinder-dry forests. Each of the three groups lobbied for wildfire protection from governments, federal and provincial, and each accused the other two of wanton carelessness that resulted in the igniting of the forests.

The wave of wildfires at the dawn of the twentieth century forced the lumber companies to consider cooperative fire protection for their holdings. Ellwood Wilson, an American who had studied forestry in Europe, was appointed in 1907 as the chief forester at the Laurentide Lumber Company mill in Grand-Mère, Quebec. Regarded as the father of Canadian forestry, he began the St. Maurice Forest Protective Association, the first grouping of Canadian forestry companies intent on protecting their resources against fires.

The "Big Burn of 1910" was a harbinger of what was to come in Canada the following year. In the hot and dry summer of 1911, the gold-rush shantytown of Porcupine in northern Ontario was inundated with prospectors, storekeepers, gamblers, prostitutes, and adventure seekers. On July 11, a wildfire formed a horseshoe-shaped front thirty-five kilometres wide around the camps, igniting stored explosives and cremating all within the town. Because of the town's transient population and lack of personal identification, no one would ever know how many perished. The provincial government, unprepared for such disasters, did nothing to fight the Porcupine Fire or help its victims — only the churches and the Eaton's department store in Toronto sent aid to those made homeless.

It was the wildfire that swept through Black River-Matheson, Ontario, on July 29, 1916, killing 223 townsfolk, that forced the provincial government — at the urging of the lumber industry — to finally draw up regulations on forest fire detection.

The problem with wildfire management in Canada was the result of mixed jurisdiction. The 1867 *British North America Act* had given Ontario and Quebec control of their forests, but when Alberta and Saskatchewan became provinces in 1905, the federal government retained ownership of their natural resources, including oil, gas, coal, and the forests. Forest fires in the Prairies were the responsibility of the Dominion Forestry Branch (DFB) in faraway Ottawa. Their agents valiantly patrolled the Slave and Athabasca rivers. But there were no regulations governing the disposal of logging slash or the indiscriminate use of fire to clear fields of brush. In 1919, to get the federal government's attention, a delegation was sent to Ottawa to discuss the situation. While they were en route, a firestorm swept through the nearby boreal forests, halting the train they were on. It ravaged two million hectares, and in its path were homesteads, hunting and trapping camps, timber berths and lumber camps, and communities, including the village of Lac La Biche.

Towns in those days were built entirely of wood — homes, schools, and shops; even the sidewalks were planks:

> When the fire was still a mile and a half away, the flames carried over and set the town afire. At that time the heat was so intense it was searing their faces. It came on first towards the church and then as suddenly as a miracle, the wind changed, and the church and priest's house were saved, and the fire raged on to the little town. The women picked up their children and ran for the lake and there the men kept them covered with wet blankets. Nothing was saved but their lives, absolutely no bit of furniture,

> no money, clothes, or food, they simply had to fight for their lives.[8]

The images — the burned dwellings, families losing everything, the blackened faces of weary firefighters, officials searching for the cause (typically, some third party was blamed for the burn) — repeated themselves with sad regularity throughout North America. The hope was that perhaps with the twentieth century's latest technology, wildfires would be contained.

CHAPTER 2

Getting a Bird's-Eye View: Early Attempts at Aerial Fire Detection

IN 1909, SIX YEARS AFTER THE WRIGHT BROTHERS' FIRST FLIGHT, AERIAL patrols over national forests in the United States were proposed. William Cox, a former employee of the U.S. Forest Service, had witnessed the flight of a Wright Flyer and reported on its potential for detecting forest fires. At the annual meeting of forest supervisors, Cox's account was "read with interest and due consideration would be given to the use of aeroplanes in fire patrol"[1] (which sounds a lot like a bureaucratic death knell).

It took Logan "Jack" Vilas, a wealthy Chicago sportsman, to demonstrate the viability of aircraft in wildfire detection. The first aerial forest patrols in history began as experiments in Wisconsin in the summer of 1915. A wealthy aviation enthusiast, Vilas shipped his Curtiss H flying boat to northern Wisconsin, where he had a summer home, to conduct an experiment. In July and August, he flew surveillance missions from the forestry headquarters at Trout Lake in Boulder Junction almost daily.[2] On June 29, 1915, he even took Chief Forester Edward Griffith for a flight "to demonstrate how easy it was to spot forest fires by air."[3] Impressed, Griffith had the Wisconsin Conservation Commission appoint Vilas as a flying fire warden, the first to hold such a position in the world. "Griffith hired him, but Vilas refused pay, saying that he wanted only a salary of 'many thanks.'"[4]

American Forestry magazine featured Vilas's achievements in its September 1915 issue. "His hydroaeroplane arising from Big Trout Lake, Mr. Vilas in a few minutes can reach an altitude of 1,000 feet and from that height can survey some 200,000 acres of forested land. If he detects smoke indicating a fire in the forest, he can report in a few minutes to the State Forestry headquarters for the district, and in a very short space of time, state forest rangers can be placed along the line of the fire."

Seattle aviator W.E. Boeing, the son of a wealthy Michigan lumber merchant, having begun the Pacific Airplane Company in 1916 and built a single-engine biplane on floats, looked to sell it. Aiming for the Canadian market, Boeing wrote in the December 1916 issue of the *Canadian Forestry Journal* the very optimistic article "Finding Fires with Aeroplanes." He then toured Portland, Oregon, and Vancouver, extolling the advantages of aerial fire patrols but made no sales in either country. Despite Boeing's lack of success, the use of aircraft in forest fire prevention began early in British Columbia.

In 1905, to manage its vast holdings of Crown land forests, the province appointed its first fire wardens. With such large territories to patrol by foot, horseback, and launches, the wardens (who were said to be difficult to find in an emergency) could do little, warning of wildfires only after they had been burning for some time. On August 1, 1908, when a wildfire swept through the city of Fernie, reducing it to ashes in ninety minutes, public demand grew for better fire detection.

Blessed with a salubrious climate (for Canada), Vancouverites prided themselves on being in the forefront of aviation. On March 26, 1910, C.K. Hamilton took off from Minoru Park, Richmond, in a Curtiss biplane, making Vancouver the first Canadian city to have an aircraft fly over it. Two years later, William Stark and his wife — the first passenger to be carried by an aircraft in Canada — also flew from Minoru Park. From there as well, Alys McKay Bryant made the first flight by a female pilot in Canada on July 13, 1913. Was it any wonder that in 1917, Thomas Dufferin Pattullo, the provincial minister of lands, became enthused with the concept of aerial patrols for forest fire detection?

Pattullo contracted the Hoffar Motor Boat Company, a Vancouver boatbuilding firm in Coal Harbour, to build a seaplane. Years later, Jimmie Hoffar confessed that he and his brother Henry knew nothing about aerodynamics when they built the H-2 and that they were unaware of details such as the need for cross-bracing on the wings. Their "hydroaeroplane," copied from a Curtiss design in a magazine, was a two-seater biplane built in a pusher configuration — the propeller faced the rear and pushed rather than pulled the plane. Its fuselage was constructed of a layer of mahogany and a layer of Sitka spruce held together with a layer of linen glued in between. The wings were copied off the Royal Aircraft Factory S.E. 6 design: unbleached linen finished with four coats of dope and two of varnish. Its wingspan was forty-two feet (12.8 metres), the width of the wings was five feet (1.5 metres), and six feet (1.82 metres) separated the upper and lower planes. The struts were also of Sitka spruce. On the underside of the tips of the wings were two watertight metal cylinders to aid in balance and prevent the wingtips from dipping into the water. The six-cylinder, water-cooled Roberts engine was shipped from Sandusky, Ohio. Construction costs of the entire aircraft were estimated at between $7,500 and $8,000.

Initial test flights were conducted from Coal Harbour in late August 1918 by Flight Commander Capt. W.H. Mackenzie of the newly formed Royal Air Force (RAF). He pronounced it to be a "really excellent machine which will fly herself." With such an endorsement, the British Columbia Forest Service (BCFS) signed a one-year lease with an option to purchase. On the gloriously sunny afternoon of September 4, 1918, while flying above Vancouver in view of thousands of spectators (some of whom were alarmed that the Germans had arrived), the H-2's engine suddenly stopped, and the aircraft fell like a rock, its wings crumpling like paper. The pilot, twenty-three-year-old Flight Lieut. Victor A. Bishop, tried to make for Coal Harbour, but the aircraft plunged into the roof of a doctor's house at the corner of Bute and Alberni streets. Bishop survived with only facial bruises, and after being extricated from the wreck by the doctor's housekeeper, he walked out the front door to adoring crowds. "Bishop showed splendid nerve after accident," the *Vancouver Daily World* enthused. In the manner that fighter pilots are

On September 4, 1918, flying above Vancouver in view of spectators, the Hoffar hydroplane crashed, ending British Columbia's initial foray into aerial forest protection.

prone to, he modestly told the audience that he would be grateful to return to the Front, "where it was less dangerous."

The remnants of what would have been the first forestry patrol aircraft in Canada were lost to history as they were expediently seized by the Hoffar brothers and souvenir collectors. With the smoothness of politicians in every era, Pattullo, who had witnessed the flight from his office, issued a press release in which he "expressed deep regret over the loss of the plane, the cancellation of the aerial forest fire patrol program, and the swift demise of the proposed provincial air service." He promised to pay for the damages to the homeowner and contract for the delivery of another aircraft for the next fire season.

Shortly after the crash, however, a financial analysis revealed that the cost of an adequate system of forest patrol would involve at least three planes making two flights daily. The cost for this worked out to between 0.4 and 0.8 cents per hectare (between one and two cents per acre) — well above the 0.08 cents per hectare that was being spent on fire detection operations at the time. The prospect of such appropriations cooled Pattullo's enthusiasm, and the government's initial foray into aerial forest protection lapsed — a victim of the unfortunate crash and financial constraints.

The aeronautical embarrassment did not affect either the minister's future or that of the Hoffar brothers. Pattullo would be elected premier of British Columbia in 1933, and the Hoffars, paid in full for the loss of their "aircraft" (and spared any investigation of the cause of the crash), sold their company to Boeing in 1927.[5]

It would not be until July 7, 1919, that a forest fire in British Columbia would be spotted from the air. The aircraft, a Curtiss JN-4 named "Pathfinder No. 2," was over Duncan on Vancouver Island when its unknown pilot noticed a mass of smoke emanating from dense forest. He flew to the spot, circled to ascertain the nature of the fire and its location, and then landed at Duncan to report to the Forest Branch, which sent out a patrol to suppress it. By then, the value of aircraft in detecting wildfires had become accepted.

CHAPTER 3

Sentinels of the Forest Before Aviation

UNTIL AIRCRAFT, SATELLITE, AND CELLPHONE COVERAGE, EARLY WARNING of forest fires in North America depended on watchers in fire towers. Supplementing their ranger patrols, the U.S. Forest Service led in the evolution of lookout systems, with Canadian fire protection organizations close behind. At its height in 1938, there were nine thousand such towers in the United States, perched on top of hills and mountains where each summer lone lookouts watched for and reported fires. It was a lonely, slow life for those who worked as lookouts. This was occasionally spiked with excitement; a fire lookout once said: "There are two types of days in the fire tower … days you are so bored that you want to jump out the window, and days that are so hectic that you want to jump out the window."[1]

Fire towers were in widespread use in most Canadian provinces. The Ontario Forestry Branch erected eighty-two towers in 1917 alone — stations built of both steel and timber. When her husband was called up to serve in the military in 1943, Mrs. Raoul Languerand became Ontario's first female fire lookout, taking his place at Ivanhoe Lake Provincial Park. With some condescension, forest officials had high praise for the accuracy of her reports.

By 1960, Ontario's steel towers numbered 329, and their observers were considered the mainstay of the forestry service.

A 1924 report on Quebec forestry by the *Canada Lumberman* magazine declared, "The forest lands from the Gulf of St. Lawrence on the banks of both sides of the river, to the upper Ottawa, are now dotted with steel lookout towers." British Columbia had 150 fire towers in 1965, some preserved today from vandals and weather by volunteers.

On a clear day, tower operators could see up to a forty-kilometre radius, and when smoke was spotted, they took a bearing with the Osborne Firefinder (invented by W.B. Osborne to pinpoint the location of fires) and determined an approximate location. The limitation was that smoke had to rise above the canopy of dense forest to be seen. The watcher then reported the fire's coordinates to headquarters by carrier pigeon or heliograph, a signalling device using sunlight, mirrors, and Morse code.[2]

The Osborne Firefinder pinpointed an approximate location of a forest fire. Before telephones, the watcher reported it by carrier pigeon or heliograph.

The latter had numerous shortcomings — they were useless at night, smoky or hazy daytime conditions made their messages unreliable, and staff had to be trained in code — but in the absence of a telephone, heliographs were prized. As late as 1928, lookout stations in Saskatchewan reported extensive reliance on them.

By 1910, the U.S. Forest Service had constructed over six thousand miles of telephone line in the national forests, the first stage of a program that expanded in subsequent years. By the time lookouts were on the wane, radios were standard equipment.

The (perhaps unofficial) qualifications to be a wilderness post lookout in the 1940s were:

> Not blind, deaf or mute — must be able to see fires, hear the radio, respond when called
>
> Capability for extreme patience while waiting for smokes
>
> One good arm to cut wood
>
> Two good legs for hiking to a remote post
>
> Ability to keep oneself amused
>
> Tolerance for living in proximity to rodents
>
> A touch of pyromania, though only of the non-participatory variety[3]

These guardians of the forests — one of whom was the author Jack Kerouac — who spent their time in the remote, isolated forest environments, had to be self-contained and possess a sense of humour. Kerouac, who in 1956 endured sixty-three days at Washington State's Desolation Peak, had been given Air Force forms to record aircraft sightings — but as he wrote in *Desolation Angels*, used them instead to roll cigarettes.

> A pivotal role of the lookout at Desolation is to relay radio messages from field crews to headquarters in the park. These radio dispatches, pertaining to ongoing wildfires, trail work, and even rescues, are too far away for headquarters to pick up.... It could be a firefighter

> Mayday, or a mishap that needs to be passed on.... [The] dispatches provided critical information to fire crews and a low-cost alternative to the exorbitant financial costs of reconnaissance flights.[4]

Other tower operators left records of their time watching for fires. With a lot of time on their hands, a number of them crafted poems. In the April 1927 California District Newsletter, Robin Adair wrote a poem describing the lonely life of those in those towers.

"The Lookouts: Sentinels of the Woodland Empire"

Way above the forests, that are in my care,
Watching for the curling smoke — looking everywhere,
Tied onto the world below by a telephone,
High, and sometimes lonesome — living here alone,
Snow peaks on the skyline, woods and rocky ground,
The green of Alpine meadows circle me around,
Waves of mountain ranges like billows of the sea —
Seems like in the whole wide world there's not a soul but me.
Peering thru the drift of smoke, sighting thru the haze,
Blinking at the lightning on the stormy days,
Here to guard the forests from the Red Wolf's tongue
I stay until they take me down, when the fall snows come.

Women were favoured for lookout positions for their ability to adjust to isolated conditions, for their skills, and for their dependability. In 1913, Hallie Daggett became the first female lookout in the Forest Service, spending fifteen years at the Eddy Gulch fire tower in the Klamath National Forest, Yreka, California. When she was hired as a "fire guard," the term for lookouts in 1913, she was paid $840 per year. Once a week, her sister would bring mail and supplies by horse, linger for a bit, then plod off, leaving Daggett alone once more. Loneliness, which often spells the end of fire lookout careers, was not a problem for the thirty-year-old Daggett. In describing her life as a lookout, Hallie said, "I grew up with a fierce hatred of the devastating fires and welcomed the [Forest Service] force which arrived to combat them. But not until the lookout stations were installed did there come an opportunity to join what had up till then been a man's fight." When asked if she ever felt lonesome there, Daggett replied, "I never felt a moment's longing."[5]

Female lookouts proved to be more enterprising than their male counterparts. They grew vegetable gardens on the mountaintops, lacking only sufficient water to irrigate them.

Enduring the Forest Service (FS) diet for months, in 1948, a lookout at the Timber Mountain Station in the Colville National Forest, WA, composed the following poem:

I like FS biscuits;
think they're mighty fine.
One rolled off the table
and killed a pal of mine.

I like FS coffee;
think it's mighty fine.
Good for cuts and bruises
just like iodine.

I like FS corned beef;
it really is okay.
I fed it to the squirrels;
funerals are today.[6]

To the very end, die-hard lookouts claimed that they could do the job better than helicopters, which can't hover safely in thunderstorms, more precisely than aircraft, which can't easily manoeuvre in narrow valleys, and more accurately than satellites, which have been known to mistake sun-warmed rocks for fires. But rising wages for observers, the cost of maintaining or replacing obsolete structures, and the ability of aerial patrols to provide more complete information on fire location and behaviour prompted North American organizations to curtail lookout construction programs and deactivate stations. The number of operating lookouts in the United States peaked at five thousand in the 1950s, then fell as aircraft and satellite detection increased, and by 2024 the Forest

Service staffed just seventy-one lookouts in Washington and Oregon, fifty-nine in California, and fifty-two in Montana, northern Idaho, and northwest Wyoming.

CHAPTER 4

The First Use of Aircraft to Detect Forest Fires

THE IDEA THAT AERIAL OBSERVERS COULD SUPPLEMENT THE LOOKOUTS FOR forest fire warnings originally sprang from the glut of aircraft and idle service personnel after the First World War. In early 1918, when it was hoped that the Great War was winding down, it was thought that the aviation technology it had given birth to could be employed to assist in the detection of forest fires. Ellwood Wilson was not alone in looking toward aeronautics as a more effective means than, as he called them, "slow moving rangers." This would make remote areas, which could not be reached without the building of an extensive and expensive all-weather road system, readily accessible. James Bernard Harkin, the first commissioner of the Dominion Parks Branch, went even further in February 1918 with a fanciful prediction that soon it would be possible to manufacture a gas that could smother forest fires and that planes could drop gas bombs.

The U.S. Air Service's forest patrols began in California. Shortly after, they were set up in Oregon, the first on August 2, 1919, with seven Curtiss JN-4Ds from airfields in Salem and Roseburg, Oregon. The JN-4Ds ("Jennies") were replaced with more dependable De Havilland DH-4 biplanes, and because pilots needed fixed points to identify their position in relation to a fire, the Forest Service painted all lookout station roofs in the test area red and white.

Air patrol was hazardous work, as flying over deep canyons and sheer peaks was new to most fliers. In addition to emergency supplies, such as parachutes, food, and water, each Air Service pilot carried a pair of pigeons to assist in communicating with personnel on the ground. Without radio, sightings were relayed to ground crews by air-dropped messages or carrier pigeons. Military planners optimistically expected to need five trained officers to oversee the 930 carrier pigeons distributed among the forest fire patrol squadrons.

This seeming willingness to assist was not altogether altruistic; the Air Service hoped to make use of the Forest Service's needs to retain better congressional funding for itself. On April 7, 1920, however, Brigadier General William "Billy" Mitchell declared that there would be no expansion of the air patrols due to a lack of available Air Service personnel. One squadron would continue air patrols and cover the forests of California, but no new patrols would be taken for the 1920 fire season. Even these were short-lived, and in 1921, Secretary of War John Weeks formally ended all military cooperation, ordering the Forest Service to outsource patrols to local private companies.

In any case, the reviews of aircraft detecting wildfires were mixed. Aviation was so new to Forest Service personnel both in the United States and Canada that they were understandably hesitant about the use of aircraft. Pinpointing the exact location of a moving fire in a poorly mapped forest was difficult for pilots, and any reporting to ground crews on landing was no longer valid. From the Forest Service's perspective, air patrols could not substitute for "eagle-eyed individuals working in lookout stations perched on mountain tops." For a summer job fee, the watchers offered continuous monitoring, while planes could only perform a fly-over a few times a day, and unless smoke appeared when they happened to fly by, their pilots would miss it. What was more appreciated by the rangers was the Fire Weather Warning Service begun in 1923, which dispatched local fire weather data from Forest Service field stations.

When he toured aircraft factories to assess basing planes in Canadian forests, Ellwood Wilson understood the insurmountable difficulties of launching them from airfields in the wilderness. He recognized, though, that Canada is, on the other hand, home to the greatest number of lakes on the planet — and these natural "landing strips" are well distributed throughout the forest area.

It was this lack of airfields that had made the aeronautical pioneer Glenn Curtiss design an aircraft capable of lifting off from water. Some of the Wrights' Flyers had carried a canoe for flotation in case they came down in water. Why not make the water the aircraft's natural habitat? Surrounded by lakes at Hammondsport, NY, Curtiss realized that such a "flying boat" would not only be safer but would also dispense with airfields — such as they were.[1]

For flying boat designers, the First World War proved a bonus. Submarine warfare made control of the air over the country's shoreline and fleets as vital as over the Front, and flying boat patrols countered the threat effectively. As the world's preeminent flying boat designer, Curtiss was flooded with orders, especially when the United States entered the war in 1917. While the first HS-1L flying boat could carry two 180-pound depth charges, they were not powerful enough to damage a submerged U-boat.[2] At a request from the Royal Navy, Curtiss increased the HS-1L's wingspan, enlarged the vertical tail, and added a balance area to the rudder. The updated plane, dubbed the HS-2L, was able to carry two 230-pound bombs under its wings and a Lewis gun on the nose. It became the standard patrol aircraft for the U.S. Navy.

HS-2L

The workhorse HS-2L, or H-Boat, was more "winged kayak" than flying machine. Dubbed "squirrel cages" because of the network of intersecting struts and wires that braced the wings, the H-Boats looked obsolete already. The aircraft's riggers had a maze of bracing wires to keep in tune and joked that if a bird released in the midst of it all could find its way out then something wasn't quite right. The whole contraption — fuselage and wings — weighed 7,716 pounds (778 kilograms), of which 320 pounds (145 kilograms) were the wires and turnbuckles.

With a few exceptions, having fought the war in fighter planes, most H-Boat pilots were unfamiliar with flying boats. It was difficult to distinguish between water and sky when landing one on a lake's glassy surface. The trick was to land close enough to the shore and use the trees as a reference point.

Officially, the H-Boats were listed as having a speed of ninety-one miles per hour at level height and eighty-five miles per hour at 2,800 feet. But many H-Boat pilots swore that they had but one speed — sixty miles per hour. They took off at sixty miles per hour, cruised at sixty miles per hour, and stalled at sixty miles per hour.[3]

The pilot and air engineer sat beside each other, the latter on the right, both facing the huge wooden steering wheels. To carry fire rangers and freight, the aircraft were modified with two more cockpits behind the original ones. The sole navigational instrument was a compass fixed outside atop the hull and between the two crew windscreens. To use it, the pilot had to stand up, brave the wind, and peer down at the needle. It was of limited value, functioning well only in level flight, spinning in turbulence when the pilot needed it the most.

The gravity-fed fuel tank on the aircraft's upper wing was supplied from two larger tanks in the hull, the air engineer pumping the fuel up to it by hand until the aircraft was sufficiently in the slipstream and a propeller-driven pump took over. If that pump failed, he did some frantic pumping to keep the engine turning over and the aircraft in the air. Despite its long, broad wings, without power the H-Boat dropped a thousand feet for every mile. Hence the avoidance of low-level flying in order to provide a precious safety margin.

The water-cooled twelve-cylinder Packard-built Liberty engine turned a huge four-bladed windmill of a propeller. Until all cylinders fired, it had to be hand-cranked by the poor engineer — fourteen revolutions of the crank for one turn of the propeller. If the taxiing was too long, the engine overheated and boiling water erupted out of the radiator, showering the crew member below.

Engine failures brought about forced landings in, if the men were fortunate, lakes with sufficient distance to take off from again. Working on the engine sometimes through the black, invariably mosquito-infested night while the H-Boat was in the water meant negotiating its slippery hull, where if a tool or bolt (or engineer) fell off, hours of ducking and groping about in the gooey, stinking water (politely called "loon manure") followed. Being grounded also meant that no one was paid until the aircraft was in the air again.

More than once, to get over the "hump" and onto the "step," the front part of the hull, the crew would tie the aircraft to a tree on shore and then, with the Liberty engine straining at full revs, chop the rope, allowing the H-Boat to catapult off. A skilful pilot knew that with its wide hull planing on the step, the H-Boat could gain flying speed if he kicked the machine around in a 90-degree or even 180-degree turn. While the flying boat could be airborne in six hundred yards, its rate of climb after was slow and laborious. To gain precious height, the pilot circled the lake he had taken off from a few times.

Puncturing the wood and fabric hull did not take much — early in 1924, an H-Boat sank in Toronto Harbour after being punctured by a whisky bottle. The porous hull absorbed so much lake water that by the end of summer, the flying boat's payload of seven hundred pounds had been halved.

The HS-2L's fuel capacity was 153 gallons, which, depending on headwinds, gave it a range of three hundred miles. This required a network of fuel caches to be strategically placed along patrol routes. If possible, these were on islands in the middle of the lake and safe from wildfire. An aircraft out of fuel meant a long walk for its crew, typically two or three days through forest to the nearest railhead.

Underpowered and overly complicated, the HS-2L was designed to operate from well-equipped naval bases and launch into wide oceans, not to be nursed into the air from isolated lakes in the Canadian bush by two men whose lives depended on their luck and ingenuity. Despite their drawbacks, these H-Boats, the ex-submarine hunters, were destined to become Canada's first bush aircraft, remaining in use until the 1930s.

In the summer of 1918, unrestricted submarine warfare broke out in Nova Scotia. When U-boats sank a few fishing vessels, it was feared that it would be only a matter of time before the packed troopships sailing out of Halifax would be targeted too. The British suggested naval air patrols, but

Called the H-Boats, the ex-submarine hunter HS-2Ls were destined to become Canada's first aircraft for detecting forest fires.

as Canada did not possess naval aircraft, the Canadian government took the unprecedented step of allowing the United States to station a squadron of flying boats at Baker's Point, Halifax.

With the armistice, the U.S. Naval Station closed on January 7, 1919, and its twelve HS-2Ls flying boats, twenty-six Liberty engines, and four kite balloons were left behind for the Canadian government. The station was turned over to the Department of Works and the flying boats put in storage. At the same time as the United States donated the HS-2Ls, the United Kingdom offered what is known as the Imperial Gift: land-based aircraft, flying boats, airships, hangars, and other equipment used during the Great War. Overwhelmed by such bounty, the government established the Dominion Air Board on June 6, 1919, to administer civil aviation in Canada.

That first summer of peace was a heady one for aviation — on June 15, Captain John Alcock and Lt. Arthur "Ted" Brown crossed the Atlantic Ocean in a converted bomber. As with nuclear power later in the century and artificial intelligence in this one, much was expected of this technology.

"Airplanes Will Hunt for Pulp in Labrador," reported newspapers that August. A former captain in the Royal Flying Corps, Nova Scotian Daniel Owen had convinced investors as to the wealth of natural resources for development in Labrador and led an aerial expedition to survey its forests. His ship left from Boston with five pilots and four radio-equipped Curtiss JN-4s. The *New York Times* declared, "Boston Expedition Finds Much Pulp Wood Accessible for Exploitation."[4] However, nothing came of this. But it was the first use of aircraft, radio, and photography in a forest survey. Owen later stated that the fifteen thousand photographs taken from the air showed conclusively how much timber could be obtained.

As it was believed that there would never be another war, in 1919, the Canadian Air Force (CAF), soon to become the Royal Canadian Air Force (RCAF), was searching for a raison d'être. Without enemies to defend Canada from, it was mandated to focus solely on civil air operations — aerial mapping, spotting forest fires, fisheries patrols, and exploring the North. The wildfires that year in the Prairie provinces led the federal government to introduce aircraft for fire patrols. To take advantage of its lakes, flying boats were sent to Saskatchewan and Imperial Gift land-based aircraft to High River, Alberta. Rudimentary air stations were established at Vancouver (fire patrols, photography, general survey); High River, Alberta (fire patrols and reconnaissance work in Jasper Park); Victoria Beach, Manitoba (fire patrols between Lake Winnipeg and the Ontario border and along the northern ends of the greater Manitoba lakes); Ottawa and Sioux Lookout, Ontario (mapping and fire patrols); Roberval, Quebec (forestry and fire patrols); and Halifax, where the station's primary role was overhauling the HS-2L flying boats and dispatching them by air or rail to other stations.

In addition to fixed bases, in 1921, the Air Board established the Northern Ontario Mobile Unit. This was an ingenious air station on wheels. The aircraft (two HS-2Ls and one British-made Felixstowe F.3 flying boat[5]) could be flown to any large lake near the transcontinental railway, while supplies, crew quarters, and a photographic darkroom followed in railway cars. The unit's first fire patrols were flown between June and October 1921 from lakes at Sioux Lookout, Banning, Minaki, and Allan Water.[6]

Much was expected of the war surplus aircraft, and the media touted them as "Guardians in the Sky," bringing the glamour and expectations of the Air Age to the Canadian bush. J.A. Wilson, the secretary of the Dominion Air Board, believed that they heralded a great future:

> The machines available are all of them war-machines of obsolescent types. Most of the Stations are merely on the banks of rivers or lakes — without hangars, workshop accommodation, or other conveniences. If flying under these conditions can produce such satisfactory results, there is no doubt that when machines especially designed have been obtained, and when permanent structures for their repair and accommodation are available, the results will be beyond all question.[7]

In September 1920, Major Clarence MacLaurin, the superintendent of the Vancouver Air Station on Jericho Beach, had an HS-2L transported by rail from Dartmouth and reassembled at Mara Lake near Sicamous. Flights were conducted between Ashcroft and Kamloops, and records show that VIP passengers were carried, including T.D. Pattullo, the minister of lands (who must have felt some vindication for his earlier championing of the use of aircraft), forestry superintendents, and fire rangers.[8] The Railway Belt lands were mapped and access points for firefighting purposes noted.

The following year, to detect wildfires on the South Coast, air patrols from Jericho Beach were flown each Wednesday and Sunday during the summer. MacLaurin's log tells of his aircrew's efforts getting firefighters to wildfires so remote that it would have taken them days to trek to them overland. "Numerous patrols were carried out safely under impossible flying conditions, with visibility nil," with the HS-2L skimming treetops and lake surfaces, carrying party after party of firefighters, pumps, hoses, tents, and provisions through "blinding smoke and heat" in otherwise inaccessible locations in the interior and on Vancouver Island. In all, twenty-one fires were detected on these aerial patrols.[9]

The single F.3 shipped out to the Jericho Beach Station received its baptism of fire in 1922. Anticipating the wildfire season on Vancouver Island that summer, MacLaurin based the plane at Campbell River. On July 28, when a wildfire on the island was reported, it was estimated that ground travel to the site would take two days and require the construction of fourteen miles of new trail. By then, the blaze would be too large to control, and the logging camps would be incinerated. This was an ideal situation for MacLaurin to prove the F.3's usefulness.

The next morning, he left with it, carrying a gas-powered portable pump, 1,200 feet of hose, hand tools, a sixteen-man tent, a camp outfit, provisions for eighty-four days, and four firefighters. The flying boat's normal payload was 4,250 pounds, and despite being overweight at 4,895 pounds, it landed through the smoke on Buttle Lake, the men and materials to be discharged into collapsible boats to go ashore. The F.3 returned with three more men, and after an all-night effort, the wildfire was brought under control. This single operation, one official noted, would alone justify all the money and work that had been expended on the Jericho Beach Station since its establishment.

Alas, that summer proved to be the swan song for fire detection by flying boats in British Columbia. On September 11, having shut down the engine of his HS-2L due to a boiling radiator and a leaky gas line, MacLaurin was attempting to glide it to a landing. It nose-dived into the water off Spanish Banks Beach, and he drowned. His passenger, John R. Duncan, died later of his injuries, but the flight engineer, A.L. Hartridge, survived.

Disappointed by the limitations of both types of flying boats, the B.C. Forest Branch refused to fund further RCAF aerial fire patrols until more modern, reliable, and cost-effective aircraft became available. Neither the HS-2L nor the F.3 were flown after 1923.

The unexpected gift of war surplus flying boats from the Americans and British did not go unnoticed by forestry organizations in Ontario and Quebec, both of which exerted pressure on the Dominion government to allow them to make use of the aircraft. "Why should Canada wait any longer to test the efficiency of the aeroplane in forest fire protection?" lobbied the *Canadian Forestry Journal*.[10] Led by the indomitable Ellwood Wilson, the directors of the St. Maurice Forest Protective Association approached Charles C. Ballantyne, the minister of lands and forests, with the proposal of basing two of the flying boats at Lac-à-la-Tortue, near Shawinigan, QC, where the association had its timber rafts, for fire patrols over the St. Maurice Valley. Although owned by the association, the aircraft and their crews would be made available to all other forestry companies in the province. Ballantyne agreed, marking the beginning of federal government involvement in aerial forestry work in Canada. Wilson somehow managed to get $2,000 from the government and $8,000 by a special assessment from his company — all of which enabled the purchase of two of the ex-U.S. naval HS-2Ls.[11]

A former Royal Naval Air Service pilot, the twenty-three-year-old Stuart Graham and Walter (Bill) Kahre, an American engineer in Halifax, were hired to fly the Curtiss HS-2L (G-CAAC). Either fortune favours the brave or fools rush in where angels fear to tread: the seaplane flight over what was mostly land went unexpectedly well. On June 5, 1919, without a familiarization flight, the pair took off from the former U.S. Dartmouth Naval Air Station for Lac-à-la-Tortue, arriving on June 8.

Throughout the flight, Stuart's wife, Madge Graham, sat up front in what was the gunner's cockpit. An accomplished artist, she sketched the landscape below, sending her husband notes on a rigged-up miniature clothing line.[12] It was Madge who christened the flying boat *La Vigilance*, the first aircraft in Canada used exclusively for detecting forest fires.

Fortunately for posterity, Stuart Graham wrote of the flight in the *Canadian Forestry Journal*.

> "The First Flying Patrol of Forests"
>
> Stuart Graham, R.A.F., in charge of hydro-aeroplane experiments in Central Quebec for St. Maurice Forest Protective Association.
>
> From Halifax to Three Rivers in 9 ½ Hours Flying Time — Dodging Storm Clouds an Hourly Pastime
>
> Leaving the waters of Halifax harbour at 2.25 p.m. with a 10-knot south-east wind blowing, we struck a

magnetic course for Cape Blomidon and crossed the North Mountains forty-five minutes later. In the Bay of Fundy, the wind changed to a fifteen-mile north-easter, which brought a heavy fog when ten miles from St. [*sic*] John. We had been flying at an altitude of 1,000 feet, and as the fog gradually forced us downward, we sighted St. John harbour below and landed — making our time in the air two hours and eight minutes for the 145 nautical miles covered. Owing to the fog, we were obliged to spend the night in St. John. On the following day, the fog had not improved, we made a start at 12.17 and circled some time, knowing it would be clearer inland to gain altitude. Then taking a north-west course with a twenty-knot head-wind blowing we flew until we reached the International Boundary where we altered our course to north. Several times during the afternoon, we sighted storms ahead, but in each case, we easily circumvented these; until, passing over Eagle Lake, Maine, we encountered a larger storm with low clouds which forced us to land for the night. We had flown for 3 hours and 23 minutes and covered 160 nautical miles.

We were then only 38 miles from Lake Temiscouata, where we hoped to obtain a supply of gasoline, so although the clouds were still low and threatening the next day, we took the air at 11.40 and arrived at Temiscouata forty-five minutes later. Here our hearts sank as our order of gasoline had not arrived. We had been obliged to fill with second quality gasoline at St. John, but here the only thing obtainable was motorboat gasoline. But we decided to try some anyway as Rivière du Loup was only 35 miles ahead with chances of obtaining some proper gasoline. We got away at 5.5 p.m. and having the wind in our favour, reached the coast in thirty minutes. The St. Lawrence was just recovering from a storm and when we had taken an extra-large

> load of fuel aboard, we were unable to get off the water owing to the cross sea running.
>
> Sunday morning with a strong north-east wind blowing proved excellent weather to continue, so we made our takeoff at 1.5, arriving in Three Rivers, 170 miles away, in 2 hours 25 minutes. Here we were met by the president and directors of the St. Maurice Forest Protective Association and the mayor of the city, the Hon. Tessier. The mayoress presented a bouquet to Mrs. Graham, whose work as navigator the success of the flight was greatly due.[13]

The flying boat's 645-mile flight — its route took it mainly overland, with two refuelling stops — was the longest made in Canada at the time, and as Graham had carried a letter from the lieutenant governor of Nova Scotia to the premier of Quebec, it is also regarded as the first delivery of airmail in the country.[14]

After making the same flight again without mishap in the second HS-2L (G-CAAD), Graham began forest inventory, aerial mapping, and aerial photographic work, spotting the first forest fire from the air on July 7 along the banks of the Croche River near La Tuque.

Ellwood Wilson's first flight in *La Vigilance* was memorable. The Liberty engine died short of the lake, and the flying boat came to rest in a cow pasture pockmarked with tree stumps behind the Tremblay family home. With the help of Romeo Vachon, a Laurentide Air Service mechanic, the HS-2L was dismantled and towed to the lake.[15]

Although Graham proclaimed in the January 1920 issue of the *Canadian Forestry Journal* that "Flying Scouts in Forestry Have Come to Stay," and Wilson had him take each of the St. Maurice Forest Protective Association's directors up for flights, enthusiasm for the scheme had evaporated. Building watchtowers and hiring fire rangers was what everyone knew, and both were less risky and cheaper than maintaining two aircraft.

Nor did other forestry associations want the two aircraft. The Air Board had built a flying boat station at Roberval, QC, and the Quebec government not only contributed $20,000 but also assisted in the erection of permanent buildings. Following Quebec's lead, the following year the governments of British Columbia and Ontario contributed $20,000 and $15,000, respectively, toward the Air Board's flying operations. Wilson's vision of forestry associations owning wildfire detection aircraft had been eclipsed by history.

La Vigilance

The HS-2L flying boat that began it all survived, or at least its bones did, to embrace another day. On September 2, 1922, pilot Don Foss and mechanic Jack Caldwell set out in the H-Boat, now owned by Laurentide Air Service, flying from Remi Lake in northern Ontario with a supply of gasoline to be delivered to Lac Pierre, a ninety-minute distance by air.

The gasoline was unloaded by 10:00 a.m., and they headed back to Remi Lake. When a severe summer storm pelted the aircraft, Foss decided to set it down and wait it out. He landed in a small lake, about a half mile long and five hundred yards wide. While waiting for the weather to clear, he realized that the lake would be too short for a safe takeoff and that they would not clear the trees at the water's edge. When the storm ended, he first cruised the lake for floating logs and then prepared for takeoff. Foss concentrated on clearing the trees along the shore, but a wingtip hit the water, and the flying boat cartwheeled into the lake. Caldwell was thrown from the plane and landed on the wing. Foss was unconscious in the submerged cockpit, and Caldwell pulled him ashore, where he regained consciousness.

The men walked along the nearby Groundhog River in search of assistance. A trapper found them, and they sheltered that night in his cabin. The next day, he led them to Fauquier, the nearest railway stop. Two days later, they returned to the scene of the crash and determined the stricken H-Boat was a write-off. The Liberty engine was retrieved but beyond repair; it was later scrapped.

There *La Vigilance* remained, slowly sinking below the water's surface, settling in the silt, until the late 1960s. It was located by a Kapuskasing businessman, Don Campbell, who alerted the authorities. Since no other HS-2Ls were in existence, the decision was made in 1968 to retrieve the plane from what is now called Foss Lake and, if possible, reconstruct it. No one was aware of the plane's history until the salvaging progressed. That it was Canada's first fire detection aircraft made it a national treasure. The restoration took place from 1970 until 1986, and the reconstructed aircraft, built with parts from three different HS-2Ls, is the crown jewel of the Canada Space and Aviation Museum in Ottawa, with the original hull of *La Vigilance* preserved separately next to it.

Wildfire reporting by aircraft in Ontario began in 1919 when the provincial forestry service gave Laurentide Air Service (formerly part of Wilson's St. Maurice Forest Protective Association) an exclusive contract to conduct fire patrols from Sioux Lookout, allowing that company to stave off bankruptcy that year.

In the summer of 1921, pilot Reg Johnston saw a wildfire while over Cliff Lake, where Wabakimi Provincial Park is today. When he returned to Sioux Lookout to report it, it was decided that he should take a ranger with firefighting equipment with him back to the fire. The ranger managed to extinguish the fire before it got out of control, the first instance of aerial fire detection leading to wildfire suppression in Canada. Despite that success, Laurentide would not continue providing aerial firefighting support for very long. This was not because such services were no longer needed; rather, it was caused by a significant increase in the need for them.

One of the worst wildfires in Ontario's history occurred on October 4, 1922, killing forty-three people and destroying the villages of Thornloe, Pearson, and Uno Park in the Temiskaming district. Much of the prosperous mining town of Haileybury burned, and thousands of its inhabitants were left homeless. In the extreme heat, every glass window in the town exploded, kegs of nails and the terrazzo floor of banks melted, and farm animals died in the fields with scorched lungs. Only a providential heavy snowfall the next day extinguished the fire.

As with the "Big Burn of 1910" in the United States, newspaper coverage of the Haileybury Fire warned the government and forestry companies that with the dry summers, the number and ferocity of wildfires was only going to increase. The next year, over eighty thousand hectares burned in Ontario alone, of which 28 percent was valuable timber. Recognition of the importance of the forestry industry to Ontario's economy forced the provincial government to shift its emphasis from overseeing the exploitation of this resource to its management.

The summer of 1924 was the driest on record, and across North America, wildfires proliferated. Firefighter (and poet) R.W. "Bummer" Ayres summed it up with bit of doggerel.

"The Summer of Twenty-Four"

Smoke and dust, fever and sweat,
The damndest season I've put in yet.
All you can hear, or think, or do,
Is fighting fire the season through.
All other work has gone to pot,
Our working plans are completely "shot."
(Suffering cats, will it never rain?)
My heart has a knock, my nerves are frayed,
My stomach's gone; my feet are splayed.
My eyes are dimmed from the backfire smoke,
My lungs are sore, and my back is broke...
A holiday for me would be.
On a southern isle in a balmy sea,
Where I could sleep, and eat and shave,
And bathe myself in the purple wave.
In its tropical rains with its glad downpour,
I'd dream of the summer of twenty-four.[16]

With the Haileybury Fire devastation still reverberating through the province, it was an opportune time on January 16, 1924, for Lands and Forests Minister James W. Lyons to lobby his Cabinet colleagues. In the minister's opinion, the province should purchase the necessary flying boats to create its own forestry aerial organization for the expected wildfire summers. Laurentide immediately offered to sell its H-Boats with airworthiness certificates for $6,000 per machine, to be delivered to either Toronto or Sudbury. In a first for any government anywhere, the Province of Ontario was about to create its own forest patrol service.

The deputy minister of national defence, G.J. Desbarats, cautioned against buying the obsolete HS-2Ls, warning that the purchase of the aircraft was not wise. He was not alone in his doubts. At the annual meeting of the Canadian Society of Forest Engineers that year, Wilson led a committee to consider aircraft requirements for Canadian forestry operations. Its recommendations were given to the Department of National Defence and taken up by Canadian Vickers of Longueuil, QC, for the design and construction of such an aircraft.

Vickers Vedette

Named for female cabaret artists of the day, the nimble Vickers Vedette made the HS-2L look like a blowsy dowager who had seen better days.

The first Vedette was completed on May 10, 1926, and reserved for Ellwood Wilson's Fairchild Aerial Surveys of Canada. Popular with its aircrews, the little flying boat was not only the first mass-produced aircraft to be both designed and built in Canada but was also the first specifically for forestry survey and fire detection. Before production ended in 1930, sixty

The Vickers Vedette was not only the first mass-produced aircraft to be designed and built in Canada but was also the first specifically made for fire detection.

Vedettes were built, of which the RCAF purchased forty-five for photographic and forest patrol duties across Canada.

Perched on top of the tallest tower of the parliamentary precinct in Ottawa is a weathervane with a Vickers Vedette on it.[17]

Why was it accorded this honour? In 1929, the foundations were laid for the Confederation Building to house the Department of Agriculture. Canada was an agricultural country, and the department spared no expense in the construction of its new home. When it was completed, the architects looked for a suitable weathervane to crown the building's tallest tower. "The Department of Public Works was casting around for something typical-symbolical-distinctive to finish the tower," wrote a reporter in the *Ottawa Citizen* on February 27, 1931, "and happening to view a moving picture show of aerial craft, the designer saw flashed across the screen exactly what he wanted as a model for a weathervane." Aware of what the country owed the little aircraft, the architects decided to immortalize it as the weathervane. Only the Canadian flag on the Peace Tower flies higher.

Ontario Provincial Air Services

In Ontario, with an eye on the immediate fire season, Lyons ensured that Cabinet rubberstamped the legalities to create the Ontario Provincial Air Services (OPAS), the world's first government aerial fire management organization. Thirteen H-Boats from Laurentide Air Service were bought at the agreed price. To cover the 2,259,725 square miles (585,266,088 hectares) of northern Ontario, the OPAS flying activities would be divided into eastern and western districts, with Lake Nipigon as the dividing point. Major G.A. "Tommy" Thompson would operate the Eastern District from Lake Ramsay, Sudbury, and patrol up to the Quebec border, while the Western District under Captain J.R. Ross based at Sioux Lookout would patrol as far as the Manitoba border.[18]

Appointed OPAS director, Capt. W. Roy Maxwell, a former Royal Flying Corps instructor, chose Sault Ste. Marie (the "Soo") as its main base, listing his reasons: the city's proximity to the air service's summer bases in Sioux Lookout and Sudbury would minimize the amount of cross-country flying and the early opening of the St. Mary's River would enable personnel to ready the planes in the water sooner than at other locations. Construction of the hangars was begun in 1924 and completed the following year.[19]

With Laurentide closed, Maxwell had little trouble persuading several of its pilots and engineers to join the OPAS. The list reads like a Who's Who of Canada's Aviation Hall of Fame: T.M. (Pat) Reid, Romeo Vachon, Al Cheeseman, C.J. (Doc) Clayton, Terry Tully, Fred J. Stevenson, James V. Medcalf, Pat J. Moloney, C.A. (Duke) Schiller, R. Carter Guest, H.A. (Doc) Oaks, Leigh Brintnell, Tom Woodside, and Frank MacDougall.

Pilots were paid $250 a month plus $2 per hour flying time. Air engineers graduated into first-, second-, and third-class ratings, receiving, per month, $185, $150, $125, respectively, plus a bonus of $1 per hour flying time. Air crews were responsible for obtaining their own flying suits, known as Sidcot suits. (In the winter of 1916, when his squadron's pilots were landing paralyzed by the cold, Australian pilot Sidney Cotton designed a one-piece flying suit that had a khaki twill proofed cotton outer lining with a rubberized cotton interlining and fur collar. The Sidcot suit was continually redeveloped and used in the Second World War.)

Answerable only to Lyons, Maxwell ran the Air Service with an iron hand, deciding who to hire and what aircraft to purchase. He was strict with the air crews and careful with the OPAS finances, once criticizing a pilot for spending $2 on a taxi ride when it should have cost $1. When Thomson asked if he could spend $18 for an iron winter stove, Maxwell sent him a $4 sheet metal one. When a rainy season kept crews indoors and grating on each other's nerves, he sent them cases of boxing gloves to use up some energy. This must have improved morale beyond expectations, as the next request was for three harmonicas.

Before radio, OPAS pilots communicated with the bases by carrier pigeon, as the RAF and RCAF did until 1943. Maxwell even had two specially bred homing pigeons imported at great expense from Europe for the purpose. However, given a pigeon's chances of surviving over a forest of patrolling hawks, the pilot's best chance was landing near a railway communication line where he could hook up his portable telephone and bellow his location to the dispatcher. Hawks weren't the only threat the pigeons faced.

There was a rumour that a pilot forced down somewhere without emergency rations had realized that even if the pigeon made it back, he was still going to be starving and made the decision that a bird in the hand … He cooked a squab dinner, officially listing the pigeon as missing.

The first ground-to-air radio communication in fire detection in Canada was from the Laurentide radio station at the Nickel Range Hotel in Sudbury.[20] When the OPAS took over the station, two of the H-Boats were equipped with wireless telegraphs that were powered by wind-driven generators. These were later replaced by storage batteries. The first use of air-to-ground radio in Canada took place that season when Monty Baker, the air engineer in an H-Boat flown by "Tommy" Thompson, spotted a wildfire in Scotia Township in the Sudbury District. He tapped all information about the blaze on his kneeboard transmitting key to the OPAS Sudbury office.

At the end of the 1925 flying season, Maxwell and Lyons could congratulate themselves. In fire detection patrols, the Air Service had flown 2,738 hours and 37 minutes in 1,312 flights. The lowest tender to do the flying had been $95 per hour, and the OPAS had been able to do it for $47.75 per hour. Fire suppression ground crews had been carried, and much of the Ontario wilderness had also been sketched and photographed.

• • •

In the Prairie provinces, each RCAF base had two Vedette patrol planes and a larger Vickers Viking. The patrol aircraft spotted and reported fires, while the Viking carried rangers to suppress them. In Manitoba and Saskatchewan, some of the aircraft had radios, but few messages were ever received. More reliable were those dropped in bags to rangers, and they in turn communicated with the pilots by laying out signal strips — a method from the First World War.

It was a pilot based out of Lac du Bonnet who made RCAF history in late August 1929. Sgt. J.M. Ready took off in a De Havilland Moth bound for fire patrols at Gordon Lake (near Nopiming Provincial Park). Haze and heavy smoke forced him to turn back to Lac du Bonnet. On route, while flying blind, Sgt. Ready lost reference to the horizon. The plane dropped into a dive and couldn't be stabilized. Sgt. Ready bailed out at five hundred feet, deploying his parachute before he and his plane ended up in the water. He swam to shore and started walking. "After five miles, Sgt. Ready arrived at Davis Lodge (located on the Bird River), where he borrowed a boat to return to base. Sergeant Ready became the first RCAF pilot to be saved by a parachute."[21]

The RCAF's role as aerial fire ranger inspired Flight Lieutenant F.V. Heakes, a future RCAF air vice marshal. His poem, "The Forest Watcher," was published in *Canadian Aviation* magazine in July 1928.

> Over lake and pine-clad forest,
> Over river flashing by,
> Sails the white-winged forest watcher
> Softly in a cloud-swept sky.
> Softly drones the distant engine,
> Over virgin timberland,
> Speaking peace unto the forest
> Where the mighty giants stand.
> Onward then o'er tracts scarce charted,
> Over cataracts asweep,
> Over mountain, plain and valley
> Over glades that lie in sleep.
> Far into the western twilight,
> Flashing wings against the sun,
> Hums the softening song of engine,
> Throbbing until day is done.

With firefighting and administrative costs soaring, Prime Minister Mackenzie King was quick to rid the federal government of the responsibility and cost of management of the federal forests in the western provinces. In 1929, transfer agreements were signed with Alberta, Saskatchewan, and Manitoba for the return of their forests, and the RCAF forestry patrols were curtailed. For a dollar each, seven Vedettes were transferred in 1932 to Manitoba for forestry patrols on the understanding that the crews hired to operate them would be ex-RCAF personnel. Under a similar arrangement, five aircraft were also transferred to Saskatchewan, and that province took over its own forest fire patrols in 1933. Based at Ladder Lake in the Big River area of that province, the Vedettes were no bargain. Only three were in actual flying condition, and two required servicing to make them airworthy. On May 27, 1936, when Vedette (CF-SAE) crashed north of Delaronde Lake after only seven hours of flight time, the remaining aircraft were grounded.[22]

By 1929, the iconic H-Boats had deteriorated beyond reliability. The last one would take to the skies in 1932, its longevity in arduous bush conditions a testimony to the pilots and air engineers that had flown and maintained it. The Vedettes were by then instructional airframes, and the last would be written off in 1940. In 1979, the H-Boat and the Vedette would be given their due and featured on a postage stamp series by Robert Bradford, curator at what is today the Canadian Aviation and Space Museum, Ottawa.

As replacements for the H-Boats, the OPAS purchased its first pontoon aircraft, which could be fitted with skis in the winter. They were four all-metal Hamilton planes and fourteen De Havilland Moths with welded steel tubing fuselages. It was Maxwell's keen interest in the Moths that influenced the British aircraft manufacturer to open a plant at Downsview, Toronto, in April 1929, for the Canadian market.

By 1930, making do with several types of aircraft, the service lacked a general-purpose type ideally suited to its requirements. It needed to have a payload of 850 pounds (the weight of a firefighting crew and its equipment) and the versatility to operate on floats in the summer and skis in the winter. The Buhl CA-6 Air Sedan, featured in the unfortunate Dole Air Race, had a short takeoff run that allowed for operation out of small, otherwise inaccessible lakes. In 1936, the OPAS assembled four CA-6s with Canadian Vickers floats at Sudbury according to their own specifications.

Frank A. MacDougall, the superintendent of Algonquin Park (and later Ontario's deputy minister of lands and forests) had the OPAS purchase an open-cockpit Fairchild KR-34C biplane built in Montreal in 1931. With it, he monitored the movement of wildlife, the activities of poachers, and the progress of campers along its waterways in the park.[23]

Whatever aircraft the OPAS bought, all arrived in time to be pressed into the worst firefighting season until then. The climatic conditions across

North America in 1936 were responsible for the "Dust Bowl" in the Great Plains and prairies. In one of the hottest Canadian summers on record, it unavoidably led to a record-breaking fire season. Beginning July 5, a fortnight of heat waves combined with lightning to create wildfires from southern Saskatchewan to the Ottawa Valley. More than 2,264 wildfires were fought in Ontario alone — and thanks to the red-eyed, weary OPAS pilots who shuttled men and supplies to them, most did not reach an area of more than a quarter of an acre. But an estimated 1,180 people died, and more than five hundred thousand hectares were laid waste that year.

Katherine Stinson was one of the earliest female aviators and in 1918 had flown the first airmail between Calgary and Edmonton. Her brother, Eddie Stinson, inspired by Lindbergh's Ryan monoplane, began building sturdy, radial-engine bush planes in the 1930s. His greatest success was the Stinson Reliant, which could accommodate a pilot and four passengers. It was nicknamed the "gull-wing" due to its elegant, tapered wings and was popular with passengers for its leather-covered interior.[24] The fuselage, tail surfaces, and wings were of welded steel-tube construction, and the whole framework was covered with cotton fabric. The Stinson Reliant 9FA was designed for the Canadian bush market, and the OPAS purchased ten in 1937, using them until 1950. Not only did it make the Air Service no longer dependent on finding lakes, but the Stinsons were the first OPAS aircraft equipped with two-way, voice-to-voice radio, which enabled aircrews to talk to ground station operators.

While the OPAS still continued to purchase new aircraft throughout the 1930s, the Depression generally crippled expenditure on fire protection agencies in all other provinces. Each reduced the number of rangers, slashed their wages, curtailed purchases to replace aging tools and machinery, and simply allowed fires in the North to burn themselves out. For fire detection, governments concluded that a system of lookout towers, linked by telephone — and after 1936 by radio — provided a more cost-effective approach than airplane patrol.

In 1944, with the easing of wartime aircraft-buying restrictions by the Department of Munitions and Supply, the OPAS received its first Noorduyn Norseman. The rugged bush plane, first flown in 1935, got a windfall with the war, when the U.S. military bought 759 of the aircraft, flying them in Europe, Alaska, the Arctic, and Antarctic. With a welded, chrome molybdenum steel-tube frame and a 550-hp Wasp engine able to lift a payload of 1,800 pounds, the Norseman remained in production until 1959, when the Canadian Car and Foundry in Fort William, Ontario, built its own Mk. V version.

With many of the OPAS aircrew resigning to join the RCAF in 1940, it was fortunate that wildfire activity in Canada decreased in the war years, influenced by a mild climatic period. The threat then was no longer lightning or careless campers but rather Japanese incendiary balloons.

• • •

Having nursed the OPAS through its formative years, the controversial Maxwell was not to see its maturity. When the Liberals were elected in 1934, he incurred the wrath of Ontario premier "Mitch" Hepburn and became the target of a royal commission established to investigate his conduct, as he had been accused of financial mismanagement. Hepburn replaced him, rewarding his personal chauffeur, G.E. Ponsford, with the position. Although no evidence of impropriety was found, Maxwell resigned from the OPAS in 1935. His career may have ended in a cloud of suspicion, but he left an enduring legacy, and he remains celebrated as the father of aerial forest protection in Canada.

CHAPTER 5

Smokejumping

IN THE GREAT DEPRESSION, DISTANT WILDFIRES DID NOT CLAIM NATIONAL attention as much as finding a job did. Unemployed men were starting costly fires all over the country in hopes of getting a job fighting them. The fire seasons erupted as the worst in the brief history of the California Division of Forestry. The division countered this rash of incendiary fires by establishing "sit tight" fire crews and stating that no pickup firefighters would be hired. But the fires continued — until the interests of the state's Board of Forestry, the Department of Finance, and the Department of Social Welfare merged. Hiring desperate men, each paid forty cents a day, conveniently solved the issue for all three departments. What could possibly go wrong?

On October 3, 1933, a wildfire that swept through Griffith Park, Los Angeles, had a profound effect on the United States. What began as a small brush fire overwhelmed the 1,370 untrained, leaderless, confused men fighting it with only shovels.

"You could tell the progress of the fire by the screams," reported a survivor. "The flames would catch a man and his screams would reach an awful pitch. Then there would be an awful silence."[1]

Twenty-nine workers caught in Mineral Wells Canyon died, and 150 others were injured. The Tillamook Burn, the conflagration that swept through western Oregon's Coast Range that same year, incinerated over 300,000 acres (121,000 hectares). But an actual forest fire in the heart of Los Angeles, within walking distance of glamorous Hollywood and Vine, which could have jumped into the celebrity neighbourhood of Hollywood Hills, attracted maximum newspaper and newsreel attention. As with the "Big Burn of 1910," the extensive media coverage had the effect of the federal government elevating fire suppression to a national priority.

The year before, T.V. Pearson, the U.S. regional forester, had written to the Goodyear Zeppelin Corporation to inquire about obtaining a Goodyear airship to carry men and equipment into the backcountry to fight wildfires. With most of the national forests roadless, flying in men and supplies to contain a fire as expediently as possible was a clean answer to a dirty problem. "Transporting firefighters by dirigibles from which ladders can be lowered to the ground"[2] was cheaper than using the strings of mules then being used to carry tools, tents, and food to where the fire crews were working.[3]

Using airships (filled with noncombustible helium) in the 1930s was not as far-fetched as it sounded. Blimps had been put into service in 1921 for fire patrol flights in California. The Los Angeles County Fire Service and Forest Service had used them extensively over the Angeles National Forest that year. It was imaginatively hoped that they could linger over a fire longer than aircraft and even carry thousands of gallons of water.

In a very candid response, Goodyear pointed out to Pearson that dirigibles could not operate in intense winds or mountainous regions and were unable to carry more than 1,200 pounds.

In 1922, the supervisor of the Los Padres National Forest in California had tried an autogiro (an aircraft powered by a conventional propeller, with unpowered rotating horizontal wings to keep it aloft) for fire reconnaissance; it looked promising.[4] Earl Loveridge of the Washington office discussed the new craft with the Bureau of Aeronautics, and the Kellett Autogiro Company in San Francisco provided an autogiro for testing.[5] It could fly at very low speeds, twenty-five miles per hour, and could land in a very small field. It could not hover motionless, although it could hold a horizontal position while it descended slowly. Convinced as to its use, in August

1938 the chief forester proposed to acquire eight autogiros. When Congress appropriated $300,000 for the development of rotary-winged aircraft the following year, the U.S. Army abandoned the autogiro for the helicopter, and the Forest Service followed suit.

Undaunted by the refusal of airships to transport firefighters, in 1934, Pearson lobbied for parachuting firefighters, even arranging for a demonstration with a professional parachutist. His idea was considered a harebrained scheme, parachutists being regarded (and with some justification) as crackpots, publicity-loving daredevils, or just plain crazy.[6]

What forced the Forest Service to consider aerial suppression of wildfires was the fire season of 1934, when the Pete King and McLendon Butte fires burned 240,000 acres of the Selway National Forest in northern Idaho.

To suppress both wildfires, more than five thousand men were employed, distributed at seventy-four different fire camps, from which they built 410 miles of fire line, largely by hand. More than one hundred trucks and travel cars were driven a total of 233,000 miles to supply fire camps with tools, food, bedding, and other supplies. Four hundred and seventy-five head of pack stock transported the supplies from the ends of the roads to the fire camps in the back country. The total cost was a little over a million dollars, not to mention the timber, wildlife, watershed, and other values that were destroyed.[7]

Not only were two lives lost, but such an expense during the Depression made the Forest Service consider fire suppression by means other than ground-travelling smoke chasers.[8] Aerial detection would supplement and, in some cases, replace ground observers in the fire towers, and suitable aircraft could deliver equipment and supplies to firefighters. The eternal mantra that the quicker a fire is controlled, the lower the human and financial cost, was never more relevant.

This was tragically emphasized on August 21, 1937, when the Blackwater Fire in Shoshone National Forest, near Cody, Wyoming, killed fifteen firefighters and injured thirty-eight. The government investigation concluded that had crews been rushed in as soon as the fire was detected, it would have been contained and the deaths averted.

Two years later (and sixteen years after the OPAS received its first HS-2L), the U.S. Forest Service took delivery of its first aircraft, a Stinson

Reliant. Although bought for forest patrols, the Reliant helped the agency explore the idea of airdrops of men and supplies, thus bypassing the need for lengthy overland trips by mules to the burn sites.

Unfortunately, the Service was unable to find a pilot willing to fly the plane on a seasonal basis. The Stinson gathered dust in an Oakland Airport hangar until the Washington office finally traded it for four Piper Cubs, two of which were sent to California. Andy Brenneis, supervisor of the Trinity Forest and an enthusiastic flyer, asked to try one out over the forest. He flew the underpowered aircraft around Hayfork and Weaverville and returned it to Oakland, but he didn't think it was substantial enough to fly safely in mountainous terrain.

The Forest Service's Aerial Fire Control Project began in December 1935, setting up headquarters at Oakland Airport. The goal of the project was to find ways and means to directly attack forest fires from the air. The team considered two basic methods: the release of unconfined liquids splashed directly on the fire and the release of containers or bombs filled with water or fire retardants. The men wanted to bomb something and devoted the next two years to designing containers, bombs, bomb racks, and bomb sights. Various cardboard and tin containers were tested in airdrops at Oakland Airport in October 1936, Livermore Airport in January 1937, and at Cuyama Ranger Station, on the east side of the Los Padres National Forest, in August 1937.

It was at the latter that one firefighter witnessed a disaster he would not forget. A small fire was started by a crew at the mouth of Quatal Canyon, and the pilot flew a Noorduyn Norseman low and slow over it, dropping five- or ten-gallon tins of water that burst as they struck the ground and splashed contents in or near the fire. As it was a hot and dry summer day at Cuyama, the fire quickly recovered from the water drops and had to be repeatedly contained by the crew. Each time the crew got a foothold on the fire, the drop plane would swoop down with another load of tins. The crew would scatter, and more tins tumbled out to burst, barely affecting the fire at all. Several repeats of this scenario soon exhausted the men, giving them a rather jaundiced view of aerial attacks on fires. The plane finally landed to refuel, allowing the crew a respite. It was to be permanent, for on takeoff, the aircraft crashed. No one was hurt, but the Cuyama experiment ended water drops that year for the Forest Service.

Cargo dropping from aircraft to ground crews was first employed on actual fires in 1929, and the service began constructing remote airfields in national forests as staging areas for firefighting equipment to be flown in. Yet even this was a challenge for the elite smoke chasers who fought fires on the ground, as it still required them to pack and haul these supplies overland by mules or horses or on foot using backpacks.

In July 1938, supply drops by aircraft redeemed themselves when an intense lightning storm started hundreds of fires in the Klamath Mountains in northwestern California. Of the seven fire camps set up, two were supplied completely by aircraft, two by a combination of pack mules and aircraft, and three by mules alone. The airdrops were inhibited by a lack of folded parachutes, which became available in 1939. Of the four hired aircraft that dropped supplies, none were specially adapted for airdropping, but they served well, making three or four round-trips per day for ten days, dropping twenty-six tons of supplies, most of which were successfully executed. A few were spectacular. There were the seven-foot-long crosscut saws that broke loose over Blue Creek, oscillating end-to-end, swooping and sideslipping, before finally coming to rest in a deep side canyon. There was also a crate of eggs that landed perfectly, on the top of a Douglas fir, where they nestled until a crew cut the tree down — and salvaged one egg. But the fire crews were said to be grateful for the drops and even amazed by them.

When a forest fire in the Little Bridge Creek area of Okanogan County, Washington, was fought entirely on the ground in 1939, at a prohibitive cost of $44,000, Fire Guard Francis Lufkin thought it was expensive. So did the bureaucrats in Washington, and official approval to parachute firefighters in to curtail a wildfire was expediently received. At first, test dummies, each one the approximate weight of an average man with equipment, were dropped; these were followed by professional parachutists with special padded suits, masks, and helmets. Training sites were situated in Winthrop, Washington, and Missoula, Montana, the last enshrined today as the heart of the smokejumper legend.[9]

The final test group of five men included Forest Service's Walt Anderson, who coined the term "smokejumpers." Years later, Anderson remembered its origin:

> It came about naturally. Smokejumpers get up in the air in a hurry. From there, they can see the smoke and go directly to the fire; no hunting, or detours because of brush, rock, cliffs, swamps. When they get to the fire, the smoke tells them which way the wind is blowing so they can land where a crown fire isn't going to burn them alive while hung up in a tree. Of all the ways to get to a fire in a hurry, smokejumping tops them all. You better call that hardy firefighter SMOKEJUMPER.[10]

Lufkin and Glenn Smith, a professional parachutist, made the first jump on August 10, 1940, into Little Bridge Creek mountainous terrain, without incident. They landed in an open space in the forest near the blaze and were able to put out the fire before it spread. Lufkin described the event in a 1970 interview, shortly before his retirement from the Forest Service: "We were there probably five or six hours until a packer came in, and we had it about mopped up. So, we loaded out stuff on a couple of mules and brought it out."[11]

Four U.S. Army staff officers visited the Missoula parachute training camp in June 1940. One of them, Major William Cary Lee, had been on a tour of Europe and, impressed with the German airborne forces, thought it a concept that the U.S. Army should adopt. Lee used Forest Service techniques and ideas in organizing the first paratroop training at Fort Benning, Georgia, and became the first commanding general of the newly formed 101st "Screaming Eagles" Airborne Division.

The first smokejumpers were hired for a two-and-one-half summer month period at $193 a month plus board, and there was no overtime or hazard pay. For the 1940 fire season, where they were used, the average cost, including initial attack and follow-up expenses, was $247 per fire. Without the jumpers, the cost per fire would have soared to an estimated average cost of $3,500 per fire. The substantial financial savings of the whole program impressed the bureaucrats in Washington.

To function as a one-man fire brigade capable of putting out fires at the source, besides their silk Eagle parachutes, the smokejumpers wore special outfits designed to safeguard against injuries during a jump.

In a 1941 article in *Southern Lumberman*, Joseph N. Hessel recalled, "Each man was provided with a two-piece padded suit consisting of a pair of high-waisted, low-crotched trousers and a high-collared jacket fitting inside the trousers so as to prevent up-thrusting snags or limbs from entering the suit. Straps on the legs of the trousers fitted under the feet so as to distribute part of the opening shock to the jumper's legs. Jumpers wear a football helmet and a steel wire mask as protection to head and face."[12]

Fred Poyner writes, "Equipment for each smokejumper also included a 200-foot coil of rope he could use to lower himself to the ground in the event of a tree landing; orange streamers to signal to aircraft above; a two-way battery-powered radio; leather gloves; a knife for cutting away tangled parachute lines; and two parachutes (a 30-foot backpack canopy made by the Eagle Company and a 27-foot chest-pack canopy as a back-up). Firefighting tools and supplies were dropped separately to a jumper after he safely landed."[13]

From the beginning, Johnson Flying Service provided the aircraft and pilots to drop smokejumpers. The first was the 1928 Curtis Travel Air, which is now at the Museum of Mountain Flying, Missoula International Airport. In homage to the best backcountry pilots of the era, on its door is inscribed "Tall Timber Flying Service." A series of workhorse Ford Model 5-A-Ts followed. They carried eight smokejumpers and their equipment, the last one remaining in service until 1967.

In an excerpt from his poem "Idaho City Jumps the Fires of Hell," author and smokejumper Tom Decker aptly conveys the urgency, excitement, and emotions.

Smokejumpers: The U.S. Forest Service's Walt Anderson coined the term "smokejumpers" in 1940. In Canada, smokejumpers fought wildfires in British Columbia, Alberta, and Saskatchewan.

Down to the airfield the pickups flew
The jumpers by now knew what to do.
"Take the Beech," Smokey said, "and the Cessna Too."
Load "both up, the whole damned crew!"

"So where's the fire? Smoke, what's the rush?
Last time we went the rangers was burning brush."
"Boys, this one's different," Old Smokey said,
And he looked away, and shook his head.

"We gotta go and do our best,
Everything you know will be put to the test.
This fire is hotter than ... hotter than, well,
We've been called to put out the fires of hell!"

Two planes flew out with 20 jumpers packed in.
20 jumpers thinking about their nights of sin.

20 jumpers thinking about their ultimate fates.
20 smokejumpers dreading the pearly gates....

They knocked the fire down, they pounded it out.
The jumpers kept goin' til the last spark was out.
They whooped and they hollered; each two man crew
doused the fires from hell and it was out before two.

Smokey sat back stunned, with tears in his eyes,
looked at his jumper, said, "Boys, am I surprised!
You did it quicker than I thought you could;
You put out the fire you put it out for good!"

Montoya grinned his Southwestern best,
"Smokey, this here fire was just a test!
The jumpers said it would be a real pity
If we'd miss tonight's party in Idaho City!"[14]

Gregg Phifer (Missoula '44) writes that "by 1941 the Forest Service consolidated smokejumping in Missoula, center of Region 1, which, at that time, contained 8 million acres of roadless area. Twenty-six men served that year and were divided into three jumper squads. For the first time static lines replaced manual ripcords on refresher and fire jumps."[15] When the United States entered the Second World War, many smokejumpers left to join the military, and the workforce shortage during the summers was critical.[16] By 1942, only five veteran jumpers remained.

Fortunately, conscientious objectors from the Civilian Public Service Program (CPS) helped alleviate the shortage. The adventure of parachuting to fires attracted men from many CPS base camps who were unable to enlist in the military. With smokejumping, they were defending their country but not killing its enemies. They received five dollars a month, which was double the standard $2.50 allowance in camps. The Forest Service fed them well, housed them in old barracks, gave them a one-time clothing allowance, and the Mennonite Central Committee took care of medical insurance. "We were a healthy lot," remembered Gregg Phifer, "but naturally suffered a number of jumping and firefighting injuries, some of them serious."[17]

With the desperate hope that the Allies would have to divert resources to suppressing wildfires, between November 1944 and April 1945, the Japanese launched about nine thousand hydrogen balloons toward North America to begin wildfires in the Western United States and Canada. The War Department feared biological warfare, but the intent was more basic.

Painstakingly built by Japanese schoolgirls, the balloons each carried four incendiary bombs and one antipersonnel explosive charge. Taking advantage of high-altitude winds, now called the jet stream, they crossed the Pacific Ocean. Of the 298 recovered, one did explode, killing a family of six in Southern Oregon. The closest that a balloon came to setting a forest fire was on February 12, 1945, when three bombs fell at Riverdale, twenty miles southwest of Great Falls, Montana. Two fishermen heard a hissing sound, saw the bombs dropping — although not the balloon itself — and watched as they exploded, starting a grass fire, which they extinguished.

Detecting the balloons was a problem, as they were invisible to radar at more than twelve miles. The air defence units feared shooting them down, as that would just deliver them promptly to the very forests they were intending to destroy. But American and Canadian air force pilots did manage to shoot down nineteen balloons, including ten in one day over the Aleutian Islands on April 13, 1945.

What saved the North American forests was that the Japanese balloons had an Achilles heel. The ballast-dropping mechanism was powered by a battery encased in a protective plastic box filled with antifreeze. The antifreeze solution was too weak, and most batteries froze at high altitude, so the balloons simply descended into the Pacific with their ballast bags dragging them down.

With wartime secrecy, the accomplishments of a U.S. Army African-American airborne unit in smokejumper history were for many years little known. The 555th Parachute Infantry Battalion (nicknamed "The Triple Nickles") was training at Camp Mackall, North Carolina, in July 1944 for duty in Europe.[18] When the Battle of the Bulge ended before the 555th could be shipped out, the battalion was ordered to participate in "Operation Firefly" and was sent to Camp Pendleton, Oregon, with a detachment in Chico, California. The paratroopers travelled by train across the United

States at a time when segregation was strictly enforced. Restaurants, restrooms, railcars, and drinking fountains were unavailable to them, while German prisoners of war were allowed access to all facilities. Once at Pendleton, only a Chinese eatery was open to them.

The 555th's parachute was the new manoeuvrable type designed by civilian trainer Frank Derry. It made circling turns possible, giving them a wider and safer choice of where they landed. Through the 1945 fire season, the 555th worked on twenty-eight fires, to which fifteen were parachuted. On the morning of August 6, 1945, the battalion received a call for fifteen smokejumpers for a fire in the Umpqua National Forest in southern Oregon. Though he had not completed his smokejumper training, Pfc. Malvin L. Brown volunteered to replace the assigned medic, who was sick. The fire was on a ridge on the south side of the summit, which was filled with towering trees. Brown fell from a very tall fir tree for about 150 feet into the ravine and was believed to have been killed instantly — the first smokejumper to die in the line of duty.[19]

With the war's end, the smokejumper ranks for the 1946 fire season included a mix of returning war veterans and college students. On June 28, 1949, four smokejumpers from Missoula flew to Washington, D.C., in a Ford Trimotor flown by Bob Johnson, the air company's owner himself. Skip Stratton had never been to the capital when he was selected as one of four smokejumpers to parachute onto the Ellipse Park between the White House and the Washington Monument. The first jump east of the Mississippi River, it was to commemorate the Smokey Bear Program, but the men said it was really a demonstration for the Easterners. The jump was televised, which was how President Harry Truman watched it, even though he would have had a clear view of the event if he had stepped out onto the Executive Mansion's balcony. Stratton remembered they came in so low, they were about eye level with tourists looking out from the observation windows of the Washington Monument, which are five hundred feet up. They were waving at each other. "It was our five minutes in the limelight," he remembered. "Then we went home and got back to work."[20] The Forest Service hoped the event would generate continued support for the smokejumper program. Little did anyone guess at the catastrophe that was about to unfold.

When a wildfire was reported in the Mann Gulch in Montana's Helena National Forest around noon on August 5, 1949, the service dispatched fifteen smokejumpers from Missoula. The cargo drop did not go smoothly; the Johnson Flying Service C-47 encountered heavy turbulence at normal drop altitude and was forced to climb before dropping the remaining cargo. Firefighting gear was scattered and the crew's radio broken. But they were met by a fire guard who was a former smokejumper. Then it all went wrong.

Disoriented and confused, the men panicked, attempting to outrun the thirty-foot flames advancing on them at an estimated 660 feet per minute. Thirteen of the smokejumpers died, one of them Bill Hellman, who six weeks before had jumped in Washington. The catastrophe that was Mann Gulch influenced the future of wildfire suppression, as the Forest Service created new training techniques and implemented more safety measures.[21] Fire-resistant clothing, the preparation of safety zones, and the effectiveness of the fire shelters were some of the outcomes.

• • •

Smokejumping came to postwar Canada because the newly elected New Democratic Party government in Saskatchewan hoped to encourage a northward expansion of settlers into the forested areas of the province. A royal commission on forestry suggested protecting the northern forests from wildfire with smokejumpers, like those being used in the United States. To land men on lakes as the RCAF Vedettes had once done was no longer fast enough. What was needed was a means of dropping trained firefighters at the very outbreak. The province's fire control supervisor, E.J. Marshall, and Ansgar Aschim, the assistant supervisor, visited Missoula, and their report resulted in a decision by the provincial government to establish the Saskatchewan Smokejumpers, Canada's (and the Commonwealth's) first aerially deployed fire suppression team.

Dennis W. Kelly had trained as a parachutist, rigger, and an instructor in the Canadian Paratroop Battalion during the war. He was given charge of the first eight-man smokejumper team, which had two locations: Prince Albert and La Ronge. During spring and early summer, the men were trained and stationed at Prince Albert Airport, and as the fire season approached, moved to La Ronge. They were paid $7.50 daily and $25 per jump, which especially appealed to Indigenous veterans who couldn't find employment elsewhere.

In October 1955, at the annual meeting of the Canadian Institute of Forestry in Saskatoon, Kelly would speak of the program. This is his speech.

> For four gruelling weeks they are subjected to physical training, how to use a parachute, and to lectures and demonstrations on First-Aid, fire suppression and how to use a compass and maintain and repair all of the equipment they use. By the end of the first week, they have to be able to run six miles, and at the end of the second, tumble safely from the back of a truck travelling at 30 m.p.h.... The men have to learn something of the mechanics and design of the "silk umbrella" and of how to manipulate it. Contrary to popular opinion, a parachute will not fall vertically on a calm day, but is designed to have a forward speed of 2 1/2 m.p.h.... Before the men are sent on a fire, they must make eight jumps — three of them in the bush....
>
> The men wear heavy canvas, snag-proof suits, helmets and masks; and we are really more worried about our equipment in areas of dead snags and broken trees than about the safety of the smokejumpers. In bush-jumping operations, the men are never without an extra length of rope. This is something they occasionally need to lower themselves to the ground when they become "hung up" in a tree. The men wear two parachutes, both of which they learn to pack themselves. The one worn on the back is operated by a static line in the aircraft and is the one mainly used. The other, worn on the chest, is for emergencies only and is manually operated by means of a ripcord....

> The first aircraft we used was a Norseman float plane … we have a hole in the belly of the plane which allows the parachutist to drop freely between the floats. Any other arrangement might cause the parachute to become fouled on the tail plane.…
>
> A question has now probably arisen in your minds … how do they get out? They have to walk out — at least as far as the nearest suitable lake. When they go they have to pack with them 90–100 pounds of equipment. The position of the nearest lake is indicated to the men on the ground by the pilot, who circles his aircraft overhead before heading towards it. As he flies on the correct course he "guns" his motor once or twice and thus enables them to take a compass shot on the disappearing plane. The aircraft returns to the "rendezvous" lake on receipt of the appropriate radio message from the men.[22]

With the availability of helicopters, in 1966, the Saskatchewan Smokejumper program was abruptly cancelled.[23] In 2024, only the British Columbia Wildfire Service employed rapattack crews (sometimes called smokejumpers) in Canada. Russia has an unconfirmed number of smokejumpers, said to be more than four thousand.[24]

Fortunately for posterity, the U.S. Forest Service still had 320 smokejumpers working from seven bases, and I interviewed three, each reply recorded in its entirety.

Jonathon Fuentes

What motivates you to do this?

I have several motivations that keep me within this program, but I will try to keep it short. Within the fire community, being able to work with people that take a professional approach to forest firefighting allows me to focus on the task I am assigned during those incidents. As a younger firefighter and within other firefighting organizations, the pressure and importance of being physically fit, intellectually stimulated, and remaining calm under stress are not emphasized nearly as much. By these pillars being reinforced in the jump program, I have full trust in my coworkers to keep their and my safety the top priority. Not having to "look over my back" or "check in" on some hazard they are dealing with is something I have found less in other programs. Call it peace of mind in an otherwise not so peaceful situation.

Other motivations include the excitement of jumping out of an airplane, flying a parachute to the ground safely, hiking around in the woods of the western United States, and finally the relationships I have been able to foster within my own program and the other eight jump bases.

How do you describe your job to the public?

I am a forest firefighter who jumps out of airplanes across the western United States to assist on fires. Generally, these fires are in remote areas where it can be difficult to deliver other resources and supplies, whereas our program is self-sufficient on initial delivery. Beside helping on fires, my coworkers and I have a host of other qualifications to assist the public or forest/grasslands, including medical training, crosscut and wilderness tactics, and large fire team positions, which can include anything from planning, logistics, and specialized equipment. I would generally add that we also have great experience in working in the natural environment across the United States, not just western forests.

What does a typical day look like for you, or is there a typical day?

We rarely have a typical day because of the ever-changing nature of our job, but I will describe briefly one at the base in the summer, which includes roll call at the onset of our day if we are not off to a fire. Afterward, we set aside time for physical training (typically an hour of weights or running, team workouts), followed by a detailed weather briefing, national fire situation report, and needs across the United States for support that typically includes us. For those not off to a fire right away, typical base assignments are rolled out, which include practice jumps, parachute maintenance and repair, other sewing and manufacturing projects, as well as any improvements we can make for both the base and crew cohesion. A typical summer day will end at sunset if we have not jumped a fire and then back at it in the morning, or just about twelve hours.

Thoughts and observations in the actual jump?

I have found that everyone has their own way of dealing with a jump and how they prepare themselves mentally to leap from the aircraft. There is a lot of coordination and important discussion between the spotters and jumpers, as well as the jumpers among themselves in their jump plan and missions to tackle once on the ground. The discussion of wind and approach of the jump spot dominates the space up until a few minutes before the actual leap. I find the most interesting part of the jump is how each person finds their own sense of calm, be it repeating the steps of the jump process, to closing their eyes and taking a few deep breaths, to finding a balance between the excitement and the professional calm that is necessary to get out of the plane and into the air safely.

What do you begin doing after the gear lands?

I will finish bagging up my jump gear and getting my line gear bag ready for the day. Then I search the closest cargo box for the tools, food, and water I can put into my line gear. Just about enough to last me twenty-four hours without having to resupply. After that, whether I have a hand tool or chainsaw, I will head toward the heel of the fire to engage and find out exactly what the mission is to create a safe anchor point.

How do you get water to remote sites, and what would happen if there's a fire too far from a water source?

If there is an option, we have malleable, one- to three-hundred-gallon containers called "blivets" delivered above the fire with hose attachments to have a small water source. We will request bladder bags and hand pumps for light water use. If there is no water, we use a tactic called "dry mopping," where we churn the hot ash, dirt, and woody material to reduce the energy available to the burning material, and it eventually will go out. We also try to use the early morning or late evening, when the soil is typically cooler and holds more latent moisture that can assist with putting out the burning material. Dry mopping is typically the order of the day, although it takes a lot longer to put the fire out. Generally, the fires we fight are above any water sources, and we expect to dry mop more than have a water source.

What's the most remote site you've ever been on?

I jumped a fire in the summer of 2019 about 250 miles east of Nome, Alaska, just below the Yukon River on the Kuskokwim River to the south. The fire was along the riverbank and ran into a large, sweeping grassland full of tussocks and mosquitoes. To scan across a land that vast, open, and devoid of humanity was both breathtaking and humbling. It is hard to describe. The seven of us spent five days out there.

Being so close to the fires, are you worried about smoke inhalation and respiratory illnesses? Do you do anything to mitigate the possible health impacts?

I have my own concerns about the smoke inhalation on fires, certainly, but there are a few tricks and self-reliant tactics to avoid it as much as you can while engaging a forest fire. There is an upwind side of a fire generally, too, so it's best to stay there, where you may not have any smoke near you, as it's blowing the other direction. To be brief, though, when we are fighting forest fires, there are many more immediate concerns on securing the fire, keeping people out of harm's way, and employing the best tactics for that mission that tend to take precedent for its immediate need.

What's the longest fire you've ever been on?

I spent just over three weeks on a fire in 2013 along the Montana/Idaho border. Our tours of duty now require two days of R&R at least every twenty-one days to reset the mind and body. In terms of hours spent awake in one shift, I jumped a fire in 2016 on the Inyo National Forest, where we did not get any real sleep for the first forty-two or so hours. I cannot remember exactly how long, but my guess is forty-two to forty-five.

What's your favourite campfire recipe?

Our rookies have a tradition of creating Spam-dominated meals, which require a few special ingredients they bring along with their gear with an eye for presentation. Spam sushi delivered on a freshly cut tree bole plate and hemlock as garnish was my favourite.

What's the most challenging part of your job?

Maintaining relationships outside of work and finding balance between a job that requires so much of my time and attention while trying to live a life outside of that. Not only live that life, but try to grow as a person, partner, and friend to people that may not have any idea how your fire career acts on your persona over time. I have found in eighteen years that I am just beginning to realize that myself.

What skills — mental, emotional, and social, for example — are important for this work?

Finding camaraderie among my peers by sharing stories, working out alongside them, as well as checking in on their mental state, as well as my own, is something I remind myself of, actions I can do to keep the job in perspective. In a chaotic environment, those tiny things become valuable to trust in others and believe in them so everyone can perform their duties at the highest level. More to the emotional practices, along with faith in my brothers and sisters, is sharing vulnerabilities and personal life with them as well. These folks should know how much of your head is in the game if we are to succeed or fail, and why. It's also a good check so we can take care of each other in the ways that are necessary at the time. Outside of work, I believe it's important to share the joys and hardships of this job with loved ones so that they know who we are outside of our homes. Also, knowing when to live your civilian life and put aside the job, since we are not one-dimensional, and celebrating life differently than our titles as firefighters is beautiful. And I also recommend seeing a therapist.

What does the offseason feel like? How do you gear down after so much high-adrenaline work?

The offseason used to be something I enjoyed as a separation from my career and my personal life. But as my career has progressed, I enjoy working longer into the winter and with my coworkers. In offseasons of my earlier career,

I enjoyed travelling around the country and parts of the world or skiing all winter and visiting friends and family. But now my family is here, and the routine of having a job, sense of purpose, and maintaining relationships across the fire world can be done locally. I tend not to have as much of an offseason and do so gladly. Ramping down from this job usually includes good diet, exercise, and good rest. But I enjoy the high-adrenaline work, and it translates into my personal life as well. I recreationally parachute for a WW2 reenactment group that travels across the world, I play hockey, snowboard, enjoy live music, playing music, cross-country skiing, hunting, and camping. These things keep up the excitement enough and get me outside, which is where I like to be.

There have been several movies about smokejumpers. In your opinion, can movies and TV accurately portray the profession?

If they tried to portray my profession accurately, it would come off as slivers of excitement and hours, if not days, of waiting. Not to mention missing family and suffering annoying injuries. It would be boring, and no one would want to watch it.

More and more, is there a problem with drones at a fire?

I haven't had many encounters, but if there are drones over a fire while aircraft are being used on the fire, it poses a serious risk to the pilots, passengers, and anyone unlucky enough to be below them if there is a crash. With drones being cheaper and more accessible to the public, coupled with their interest in filming or observing fires, it is only going to become an increasing hazard.

What do you think would surprise people most about this work?

The thing that comes up most often is that we sew and manufacture all our equipment in-house, except for the parachutes themselves. We construct the jumpsuits, harnesses, deployment bags, cargo chutes, and all the straps from bolts of fabric and webbing every winter to replenish our cache. Most jumpers are FAA-certified as official parachute loft riggers. It always amuses me to see some tall, muscular bro in his gym gear toiling away at a fine detail sewing project.

Emma Hawn

What motivates you to do this?

The pride I take in being a smokejumper is what motivates me to do this job. It is a challenging job for me for which I must continually train physically and mentally, even though I have been smokejumping since 2017.

How do you describe your job to the public?

I describe my job to the public as just being a wildland firefighter first, like any other forest firefighting job. But I explain that as smokejumpers, we usually focus more on I.A. [initial attack] fires instead of the big ones that make the news because we can make it to the fire more quickly than driving and hiking to a fire and usually put it out before it gets too big. We are a national resource that can be sent anywhere in the country that needs us, we require a little more physical fitness than other fire positions, and we have more fire experience/knowledge than most other fire positions.

What does a typical day look like for you, or is there a typical day?

A typical day during the off-season consists of working out and working on sewing/manufacturing projects to build all the gear required for our job in the summer. Some people spend a lot of time working in Region 8,* helping with project work and prescribed burning.

A typical day in the summer is we come in the morning for roll call, then work out if there are no pending orders, then do a weather briefing, then catch up on parachute rigging in the loft or packing cargo boxes, then catch up on any admin stuff you might have ... all the while, the alarm could go off at any time for a fire jump.

Thoughts and observations in the actual jump?

You take into consideration the terrain, the winds, what the fire is doing, where you'll be able to have cargo dropped, how many jumpers will be needed for the fire, and if you'll be able to pull the jump off safely.

---

* The USDA Forest Service has divided the United States into 10 Regions. The Southern Region, based in Atlanta, Georgia, is R8.

What do you begin doing after the gear lands?

After the cargo lands, you are packing up all your jump gear and putting it somewhere it will be safe from the fire, taking what you need in your line gear that you will have to work on the fire for the rest of the day, food/water/tool-wise, calling dispatch if you are the IC, and then hiking to the fire, usually using Avenza maps.

How do you get water to remote sites, and what would happen if there's a fire too far from a water source?

You get drinking water from cargo drops if you are there longer than three days. As smokejumpers, we usually do not have the luxury of having water on fires unless it comes from bucket drops from a helicopter. Sometimes we can get blivets from a helicopter if it makes sense. So, if we don't have any water to fight the fire with, we just dig line and create a saw break in the fuel and then mop up once the fire is contained with the line.

What's the most remote site you've ever been on?

The most remote spots I have been in are in Alaska, in the upper Yukon, and above the Artic Circle.

Being so close to the fires, are you worried about smoke inhalation and respiratory illnesses? Do you do anything to mitigate the possible health impacts?

Yes, I worry about smoke inhalation and respiratory illnesses. I have been around people who have experienced these, and I have gotten sick from a prescribed burn where I was required to hold the line.

What's the longest fire you've ever been on?

The longest fire I have been on was twenty-one days.

What's your favourite campfire recipe?

My favourite campfire recipe is Spam sushi.

What's the most challenging part of your job?

The most challenging part of our job is the unknown. You never know how long you could be gone on a fire, how hard the jump spot will be, when the alarm will go off, where you could get sent, how intense the fire will be, or how long the pack out could be.

What skills — mental, emotional, and social, for example — are important for this work?

The most important skill is to be flexible, because there are so many unknowns. Mental toughness, physical fitness, fire knowledge, [being] fun to be around, and a positive attitude are what are important to me.

What does the offseason feel like? How do you gear down after so much high-adrenaline work?

The offseason is the best feeling, knowing how hard you worked all summer, and you finally get to go back to your normal life and routine. I ski a lot in the winter and spend time with my pets, making up for lost time. I also work out a lot in the offseason and try to spend as much time with family and friends as possible.

Aaron Thorp

What motivates you to do this?

Working outdoors and not under a roof is always a bonus. But as someone who has a forestry background, I'm attracted to the stewardship of the land aspect of the job. Protecting and managing public lands is a great honour.

How do you describe your job to the public?

I usually tell people that I fight fires for the Forest Service. If they ask for details, I'll tell them about smokejumping and explain to them that our purpose is getting to fires early and fast to try to keep them small and manageable.

What does a typical day look like for you, or is there a typical day?

There's really two typical days for a jumper. One is simply waiting for the call to a fire. While waiting for the call, we'll be filling the day with tasks that keep a jump base ready for that call. This involves packing parachutes, making food boxes, maintaining/repairing equipment, and staying physically fit, training, among many other tasks. The other typical day is being on a fire and conducting fire suppression tactics that best suit the situation.

Thoughts and observations in the actual jump?

Jump operations (ops) is like a very choreographed dance where the planned choreography can change instantly. There are specific objectives for every jumper to complete to get to the ground safely while also keeping other jumpers safe and able to get to the ground successfully. So, your thoughts are generally centred on the observational cues, like wind direction and strength, hazards in the jump spot, or the complexity of the jump. Those are just a few of many, many things a jumper is thinking about that helps them complete jump ops safely and successfully. While jumping is fun, I find it to be very rare to have enough RAM to think about having fun and think [instead] about all the correct dance moves to complete the jump safely. Only after the jump and thinking back on it do I realize how fun and exciting it can be.

What do you begin doing after the gear lands?

There are jump ops; that's when the people get to the ground. Then there are cargo operations; that's when the plane drops out all our tools and food. After that, we gather everything up and start grabbing what we will need to manage the fire and start making plans.

How do you get water to remote sites, and what would happen if there's a fire too far from a water source?

The water we carry in the cargo is only for drinking. If we are in a site that is remote enough that we can't be resupplied by vehicle, then we can call the base and request more water to be delivered by paracargo.

What's the most remote site you've ever been on?

That would be a fire I jumped in the Bob Marshall Wilderness. The jump spot was at the southernmost tip of a distinct cliff feature called the

Chinese Wall, which is over a thousand feet tall and spans twenty-two miles through the wilderness. It was an unforgettable experience. It's crazy to think that I get to fly and parachute to a place like that. A place that some people pay good money and take time off from their day to day to get to. I felt very fortunate.

Being so close to the fires, are you worried about smoke inhalation and respiratory illnesses? Do you do anything to mitigate the possible health impacts?

Just get out of the smoke. Of course, there are times where you are going to breathe smoke, but for the most part we fight fire starting at the heel of a fire, so the smoke is usually moving away from you. Again, there are exceptions, but generally you can manage it.

What's the longest fire you've ever been on?

The longest fire I was on was when I worked on a hotshot crew. We were on assignment for twenty-eight days in a remote part of Alaska. Current policy says we can't work more than twenty-one days, but the only way out was to get picked up by plane, and that could not happen because of the heavy smoke in the area.

What's your favourite campfire recipe?

That would be a Spam dessert dish that was created by a jumper named Sapp. It's cooked, sliced Spam that has a doughnut hole punched into it, which is then filled with a Rolo topped with another thin slice of Spam, which is then topped with some peanut butter and jelly sprinkled with hot coco powder dust and garnished with the mini marshmallows from the coco powder mix. It's the most amazing salty/sweet dessert ever created from a food box.

What's the most challenging part of your job?

Staying physically healthy. Firefighting is a demanding job on your body and requires you constantly maintaining a level of physical fitness that for me personally is challenging to maintain. Staying in shape for firefighting is a lot of work and time! But the job would be horrible if fitness is denied.

What skills — mental, emotional, and social, for example — are important for this work?

It's important that you can work with others, because firefighting puts you in a very close working relationship with your coworkers.

Wildland firefighting is a stressful job. You're away from your home, family, and friends a lot. That is stressful. It's important to be able to manage that stress, and everyone must find their own way to do that. Hobbies on your off time, taking time off when needed, anything that helps manage stress to keep yourself mentally sound. Everyone has a stress tank that slowly empties, and you need to make sure you're doing what you need to fill it back up again.

There have been several movies about smokejumpers. In your opinion, can movies and TV accurately portray the profession?

Not a chance. Nobody would want to watch what actual firefighting is like, jumping or otherwise, as exciting as it can be — it's just not sensational enough; it's just working out in the woods. Firefighting is "moments of greatness, hours of boredom." Most of the time it's just work; there's never a love story, certainly not romantic ones, there's no evil pyromaniacs, and when bad things do happen it has never felt right to me when people have tried to make it into a movie. Someone could certainly put an interesting documentary together.

More and more, is there a problem with drones at a fire?

Yes. I've been on fires that have whole air operations shut down because of private drones flying around. Drones are still new to the world, and we need more policies surrounding their use and for people to respect those policies. It's hard to say if it will get worse before it gets better.

What do you think would surprise people most about this work?

I'd say the part where we jump into fires. It seems there are a lot of people that don't know smokejumpers are a thing.

• • •

Eighty-four years after it began, it is likely that the smokejumping profession is sliding into obsolescence. The decrease of roadless areas in the forests, budget cuts, and base closures, extended range helicopters, and rappel crews all point to its eventual demise.[25] Whatever its future, smokejumping is a celebration of a moment in aerial firefighting history when individual gusto and gumption fulfilled Francis Lufkin's vision.

CHAPTER 6

Water Bombers: A Beginning

IF AIRPLANES COULD FLY OVER A FIRE, WHY COULDN'T THEY DROP WATER on it? On a solo circumnavigation of the planet by air in 1929, the German aviator Friedrich Karl von Koenig-Warthausen saw a wildfire when over the Santa Lucia mountains of California. In his book *Wings Around the World*, he conjectured that if an aircraft could have its bomb bay modified to hold water, it could drop its payload over a wildfire to contain it.

A similar idea had already been put into practice. Using aircraft to spread insect-killing "dust" over farmers' fields began on August 3, 1921, when U.S. Army Air Corps pilot Lieutenant John A. Macready, piloting a specially modified Curtiss JN-4 Jenny, spread lead arsenate over a grove of catalpa trees at Postmaster Harry Carver's farm in Troy, Ohio. The plane, fitted with a small makeshift hopper and a release mechanism attached to its side, flew twenty to thirty-five feet over the orchard, spreading the powdered insecticide to kill the caterpillars eating the leaves of the trees.

That a pilot could do in minutes what would have taken workers on the ground days to complete was not lost on forest agencies. Demonstrating that this could be applied to dousing forest fires, in 1931, California crop-duster pilot Charlie J. Jensen, who had provided aircraft to Howard Hughes for his *Hell's Angels* movie, installed a pair of watertight hoppers on either side of his Jenny biplane and made two water drops on a wildfire east of Oroville, California. Unfortunately, the hoppers prevented his aircraft from being flown safely at lower altitudes, which eliminated the possibility of making successful uncontained water drops.

In May 1919, the *American Forestry Journal* reported that aircraft bombing wildfires held promise. "It is believed that bombs charged with suitable chemicals can be used with good results." The U.S. Forest Service began its Aerial Fire Control Experimental Project in 1929, using a Ford Tri-Motor to drop water from a wooden beer keg over a fire. This progressed to dropping chemicals and explosives from aircraft to attack fires; monoammonium phosphate added to water as a retardant was also tested. Foaming agents were evaluated as well, but since viable airtankers remained beyond reach, none of these options could be used in a real-life situation.

The Ontario Provincial Air Services had considered making use of the province's abundant number of lakes to bomb a wildfire with water, but pumping enough into an aircraft made it impractical. The idea of a water-scooping aircraft came to one Charles "Carl" Crossley in Temagami, Ontario. "The idea at first occurred in a casual kind of way, as such inspirations often do." So wrote author Bruce West.[1] "Carl Crossley, a veteran OPAS pilot, was sitting around 'chewing the rag' one day with Pete Marchildon, the district forester at North Bay. Pete had been reading newspaper reports about the incendiary bombing of German and Japanese cities to initiate firestorms. 'You know, Carl,' he remarked, 'it would seem to me that if we put our minds to it, some way could be worked out to bomb forest fires the same way things like bridges and factories are being bombed over there in the war. Only with water, instead of explosives."[2]

Carl Crossley had joined the OPAS in 1926 after flying Felixstowe F.3s for the Royal Naval Air Service in the First World War. Employed first as a mechanic, he later flew H-Boats. In 1940, although by then fifty years old, Crossley enlisted in the RCAF and was posted to 12 Communications Squadron, Ottawa, to fly VIPs about. He finished his second war service in 1944 and returned to the OPAS.

Posted to Temagami in northeastern Ontario, Carl became interested in the idea of aircraft dropping water on forest fires. He wondered if it might be possible to fly over a hard-to-get-at wildfire and drop water on it before it became a major conflagration. The water could be scooped up into a tank under the aircraft or in its belly between the struts of the landing gear. He decided to experiment using Frank MacDougall's tiny KR-34 biplane then at Temagami.

A forty-five-gallon steel drum was installed in its front cockpit, and in its bottom, Carl welded a sixteen-inch length pipe, three-and-half inches in diameter. "The pipe extended below the fuselage and was intended as a snorkel," remembered Jack C. Dillon, the forest fire control officer. "From that protrusion with the use of elbows and nipple, he made a further extension which carried the inlet to a fixed point adjacent to the float water rudders."[3]

Crossley's idea was that water would be injected into the drum while he was taxiing at speed. Unfortunately, the little KR-34 could not taxi fast enough to generate adequate pressure to force water into the drum. Using a fire pump instead, Crossley filled the drum and looked for a wildfire to assess his theory. Firefighting history was made when he flew over a small brush fire and on the third attempt hit it with the water.[4]

MacDougall suggested that he might be better served using the aircraft's floats as water containers. With a more powerful Noorduyn Norseman (CF-OBJ) on floats, Crossley worked on the plane, installing a scoop-up gear with a release mechanism and the controls for operating the water-dropping gear by the pilot's seat. In tests, each float held fifty-five gallons of water, which was sucked up while taxiing. It took only nine seconds to jettison the water from their rear. Crossley made three successful drops on a fire on August 26, 1945, and that is taken as the date of the first recorded water drop.

This Canadian invention would revolutionize aerial fire suppression. But Crossley was not lauded for his ingenuity, having incurred the wrath of OPAS director George E. Ponsford for supposedly carrying unauthorized passengers in the Norseman. The charge was never proven. He left the OPAS in 1946 and, after flying for a few companies, retired in 1960.

Inexplicably, instead of developing Crossley's water-dropping method further, the OPAS began bombing fires with water-filled paper bags. This proved to be disastrous. When a thirty-five-pound blob of paper scored a direct hit upon a smoldering fire, it scattered the embers in all directions, starting new fires in the underbrush.

In an incident that could have had international consequences, OPAS pilot Reg Parsons accidently dropped the water bombs on homes across the U.S. border of Sault Ste. Marie.

Frank MacDougall, appointed the deputy minister of Ontario's Forests and Land Department in 1941, took his OPAS floatplane with him to Toronto, mooring it at the waterfront so he could continue locating fires. It was in a discussion with his old bush pilot friend Phillip C. Garrat, president of De Havilland Canada (DHC), that MacDougall first considered what the replacement of the Noorduyn Norseman should be — a semitransport and fire suppression aircraft.

De Havilland Canada had just hired the legendary bush pilot Clennell Haggerston "Punch" Dickins as director of sales. Contacting his network of fellow bush pilots, Dickins asked them what they wanted in a bush plane. Suggestions poured in from across Canada. It had to be all-metal, rugged, winter-ready, able to land on dirt, snow, and water, STOL (short takeoff and landing), have ample cargo space, large cargo doors (on both sides, please), a stressed floor for heavy loads, a radial air-cooled engine, low stalling speed, slotted flaps ... it was a bush pilot's Christmas list.

MacDougall promised Garrat that if his company could fulfill all of these seemingly impossible expectations, he would guarantee that the OPAS would place an initial order for twenty-five aircraft.

By happy coincidence, at that moment, De Havilland Canada's engineers were Frederick Howard Buller, Richard Duncan "Dick" Hiscocks, and the great Polish engineer "Jaki" Wsiewołod Jan Jakimiuk — all three unsurpassed in aeronautical talent. Thanks to their efforts, the first DHC-2 Beaver flew on August 16, 1947, and the first production aircraft was delivered to the Ontario Department of Lands and Forests the following year. Buying forty-five aircraft, the OPAS became the Beaver's largest civilian operator. Regarded as the best bush plane ever built, the Beaver has since attained cult status, perhaps as great as the DC-3.

In 1945, while in the RCAF, Flight Officer Tom Cooke took part in a program set up by the Ontario government for chemical spraying of the budworm-infested forests near Port Arthur with specially equipped Cansos. This may have stirred his interest in aerial suppression, since he joined the OPAS the next year and became involved in projects that contributed to the use of aircraft in wildfire control. This included development of a radio altimeter to assist in glassy-water landings and a hand-held optical device for estimating the size of forest fires as well as the distance from a fire to roads or sources of water for fire pumps.

Flying the Beaver and the larger DHC-3 Otter, in 1952, Cooke experimented with waterbombing. He considered outfitting an Otter with a large tank with a watertight door that could be opened to dump the water. The problem was that the water pressure that kept the tank closed would make it difficult to open in a fast pass over the fire. Air engineer George Gill suggested that he adapt Crossley's idea of installing tanks in the floats. The new version would have open-top tanks on each pontoon, which would roll over outward, dump their loads over the fire, and then roll back for another filling. Scoops would fill the tanks rapidly while the aircraft was taxiing.

Cooke's and Gill's ideas were put into use. The DHC Beaver was fitted with two forty-five-gallon float-mounted tanks, which were improved by reversing their roll inward toward each other, and 210-gallon tanks suspended from the bellies of the Otters. The pilot released the load by pulling a lever that in turn activated a cable-and-pulley arrangement that rotated the tanks. Testing demonstrated that the eighteen-second interval between touchdown and takeoff with a full load eliminated the time lost under the old technique; aircraft landed for the reloading of water bags. Both methods would allow the pilot to make continuous waterbombing sorties from the nearest lake in a brief time.

The OPAS enjoyed considerable success with the system on a fire in the Sudbury District that summer and equipped its Otters and Beavers for the fire seasons. In July 1957, Cooke made history flying a tank-equipped Otter to contain a mile-wide fire front in the Sudbury District. After testing revealed improved water-dropping, in 1964, the department began fitting some Otters with two-hundred-gallon detachable "belly tanks" attached to their fuselage. Appointed OPAS director, Cooke would be inducted as a member of Canada's Aviation Hall of Fame in 2004.

A lesser-known Air Service development involving Beavers was an air-to-ground system known as the ground hailer or "voice from the air." It was used to warn poachers and rescue lost hikers. Shortly after its introduction, a Beaver pilot spotted a bush settler leaving a brush fire still burning. With his ground hailer, he warned the man to put the fire out. Then, to ensure compliance, he radioed a nearby ranger station. A ranger checked at once, found no fire, and asked the settler if he had been burning brush. "Yes," exclaimed the settler. "I was burning that brush, but a voice from the sky tells me to put that fire out. So, I did it, and then I come home." The hailer enabled pilots to direct firefighters and lost persons and warn anyone endangered by fire.[5]

In December 1963, De Havilland Canada followed the Beaver with the Turbo Beaver, powered by a Canadian Pratt & Whitney/United Aircraft of Canada PT6 turboprop. Of the sixty built, twenty-five were delivered to the OPAS. Its floats once more served as water tanks, but a change was made: the doors were on the bottom and they had an increased capacity of 140 gallons.

In Canada, dotted as it is with rivers and lakes, waterbombing was practical, but in the western United States, with its mountainous terrain and few water sources, it was less so. The U.S. Forest Service had been considering aerial tankers to drop fire retardant, but they lacked sufficient carrying capacity and precision bombing equipment to hit small fires accurately.

After VJ-Day, having recently bombed German and Japanese cities to rubble, sometimes at extreme range, the United States Army Air Force (USAAF) looked to adapt to a world at peace. In November 1945, the Forest Service succeeded in getting a cooperative agreement with the USAAF to begin experimental firebombing. The effectiveness of explosive bombs, "which would throw dirt on the edge of a fire and thereby retard its spread" was considered. First at Eglin Field, Florida, and later at the Malmstrom Air Force Base, Montana, military and forestry specialists demonstrated that with modifications to its bomb racks, a B-29 Superfortress, "The Rocky Mountain Ranger," could carry eight 185-gallon tanks as water bombs. Two P-47 Thunderbolts could make glide bombing runs, dropping two 165-gallon tanks that would rip open on impact, spewing water over the fire area.[6] The Forest Service engineering shops at Missoula modified the tanks to include a tail fin, a "burster well" for an explosive charge, and a nose adapter for the fuse.

The B-29's bombs were filled with ammonium phosphate solution and fortified foam; their proximity fuses were designed to cause them to burst fifty feet above ground. That summer, in seven missions flown at three thousand feet, the Superfortress dropped forty-six tanks on test fires. Each tank's burst was seen to cover a swathe forty-eight feet wide and one hundred and eight feet long. During each test fire, officers from the Aerial Bombing Evaluation Board and Forest Service technicians made detailed measurements and observations, including the rate of spread, the state of the terrain before and after bombing, weather factors, burst height, and water dispersion pattern.

With summer's end, the tests ended as well, and the military personnel and aircraft returned to their bases. The results were said to be "inconclusive," and the proposal to deploy seventy-five fighters and thirty B-29s as fire bombers for the 1948 wildfire season was abandoned. The rationale was difficult to discern. Was it concerns about the destruction of the landscape in forests or worry that the shrapnel from exploding tanks were dangerous to ground crews and wildlife? Or so soon after Pearl Harbor, was the dropping of bombs on American soil likely to disturb some? More likely, with the start of the Cold War, a downsized military wanted its aircraft back to bomb something other than wildfires.

What did come out of this was "Operation Firestop," a multi-agency, year-long brainstorming session that took place at Camp Pendleton, California, in January 1954. Using its scientific and military might, the United States had bombed its enemies into submission. Now it was thought that it could adapt those very powers to eradicate wildfires. Techniques to drop retardants accurately from the air, the use of borate, and helicopter applications that included hose-laying, water-dropping, and being lowered to fight fires were explored. This was as much of a war as fighting Communists in Europe and Asia.

As a result of these discussions, the U.S. Forest Service rented a Bell 47-6-Z in 1957 and assigned its first operational helicopter crew to wildfire patrol. Nothing is so powerful, as the French author Victor Hugo observed, as an idea whose time has come.

CHAPTER 7

Postwar Water Bombers

FOLLOWING THE SECOND WORLD WAR, THOUSANDS OF SURPLUS AIRCRAFT became available for conversion to forest protection purposes. As in 1918, postwar legacies included the superabundance of cheap aircraft and unemployed pilots. The existence of the former presented the U.S. War Assets Administration and the Crown Assets Disposal Corporation (CADC) in Canada with complex decisions to make.

The United States alone had churned out three hundred thousand military aircraft during the war, and after VJ Day, 230,000 of them, new and used, remained. The military, looking forward to acquiring the new jet aircraft that were coming on the market, needed few of them. They had to decide which aircraft to destroy and which to sell to the civilian market. No one wanted the fuel-thirsty B-24 bombers or the high-performance P-38 Lightning fighters. There was no market for the wood-built Anson or the famed Mosquito fighter. The A-26 Invader avoided the smelter only because the U.S. military recognized it as an ideal counterinsurgency aircraft for future wars such as those that would be fought in Korea and Vietnam. The C-47s and C-54s were snapped up by the starved airlines and the sturdy Boeing Stearman and N3N biplane trainers by crop-duster companies and flying schools.

In Canada, the abundance of surplus RCAF aircraft in the postwar years led to a resurgence in aerial fire detection. The British Columbia Forest Service (BCFS), which had not used aircraft since the 1920s, rented two Fairchild 82 floatplanes for fire reconnaissance in 1945. The next year, Russ Baker's Central British Columbia Airways, soon to become Pacific Western Airlines (PWA), was contracted to provide two Cessna Cranes on wheels or floats to be based in Kamloops and Nelson. Parachute tests to drop equipment were conducted with the Cranes using war-surplus cargo chutes. In 1949, an annual contract with PWA followed for six float aircraft that included four Beavers, a Junkers 34, and a Fairchild 71. They were based at Vancouver, Lakelse Provincial Park, Prince George, Kamloops, and Nelson. On the other side of the country, to complement its lookout stations, the New Brunswick Forest Service hired a Fleet Canuck for sketching and photographing fires and contracted Maritime Central Airways for aerial patrols.

While there was limited use of aircraft postwar, for the most part fire management in Canada had not changed since the previous century. Wildfires that did not come within fifteen kilometres (nine miles) of human habitation were allowed to run their course. Those most affected by this "let burn" policy were the First Nations and Métis who lived in the bush. This was never more apparent than the Chinchaga Wildfire of 1950, which made the Guinness World Records because of its size (approximately four million acres in western Canada) and length of burning time — 222 days. As its firestorm plume of smoke moved across Canada and circled the entire planet, millions thought the Third World War had begun; it had only been five years since the mushroom clouds of Hiroshima and Nagasaki. What few fire bombers there were in 1950 were grounded by the smoke. Since it was not suppressed, the Chinchaga Wildfire remains an unnamed "ghost fire" today, remembered only by the Indigenous people whose communities were wiped out.[1]

• • •

The development of airtankers postwar would change firefighting forever. The technology was born as a result of a serendipitous accident in an event unrelated to wildfire fighting. In May 1953, the Douglas Aircraft Company flight tested its DC-7 prototype at Palm Springs Airport, east of Los Angeles. Its passenger cabin had been fitted with water ballast tanks to enable the test pilots to alter the aircraft's centre of gravity to test handling. At the end of the exercise, the aircraft made a low pass over the airport runway and dumped its ballast through three six-inch valves in the airplane's belly.

The wide, mile-long swathe of water caught the attention of observers on the ground and impressed the Los Angeles County Fire Department and the California Division of Forestry. They asked the Douglas Aircraft Company to repeat the water drop on brushfires set at a natural dry lakebed in the Mojave Desert. On this occasion, the valves released the water as an ineffective mist and the fires flared back up soon after the drop. Douglas engineers then increased the diameter of each valve to eighteen inches and demonstrated that free-fall water dumped all at once just might work. Encouraged, the California Division of Forestry asked the University of California's School of Forestry to set up instrument testing to quantify the effectiveness of what might be a new way to fight wildfires.

When the DC-7 was returned, a Grumman Avenger TBM was leased from Paul Mantz, who ran a fleet of planes used by the movie industry, to continue the tests. The Grumman Avenger TBM (for Torpedo Bomber General Motors) had not only played a key role in sinking Japanese warships in the Pacific, but in 1944, before amphibious assaults on Japanese-held islands in the Pacific, military Avengers nicknamed "Flying Flit Guns" would also spray DDT on the islands to combat malaria, typhus, and dengue fever.

Powered by a massive Wright radial, eighteen-cylinder R2600, 1,950 horsepower engine, the Avenger was an ugly, bloated aircraft. Originally built for a crew of three — a pilot, navigator, and gunner — its overall ruggedness and stability were legendary, and pilots said it flew like a truck. Called "the Turkey," it lent itself to firebombing because Grumman had designed the aircraft around a large bomb bay, allowing for the Bliss-Leavitt Mark 13 torpedo, a single two-thousand-pound bomb, or up to four five-hundred-pound bombs.

Mantz built a plywood tank into the Avenger's torpedo bay for the slurry, and when it leaked, he replaced it with a Weather Bureau balloon. When the torpedo bay doors were opened, the balloon burst and dumped the contents. Later, Mantz installed a 170-gallon metal tank in the aircraft,[2] and on September 1, 1954, it made two drops on the Jamison fire near Lake Elsinore. Firefighting researchers then had Mantz fly his TBM over the deep canyons at U.S. Marine Corps Camp Pendleton, making water drops during different wind conditions, each of them analyzed for dispersal, wind drift, effectiveness, and other criteria.

Regular use of planes for the waterbombing of forest fires followed a tragedy that occurred in 1953 in Mendocino National Forest. At 10:00 p.m. on July 9, 1953, a fire broke out in a canyon. A team of firefighters was tackling another fire nearby. Taking a break from battling that fire, they turned their attention to the Mendocino fire. They were operating in the canyon when the wind suddenly kicked up from the opposite direction, creating a rapid flare-up in the thick chaparral brush. Nine of them used a rope to ascend the steep canyon walls while the others ran down the slope. The flames raced down the canyon at fifteen miles per hour, overtaking and killing fifteen trapped men.

The fatalities prompted California foresters to look for better ways to fight fires. At the time, Joe Ely, the Mendocino National Forest fire control officer, was fighting a fire in Angeles National Forest. According to his son Frank, Joe became passionate about making firefighting safer for his men. Ely had the idea of using crop-duster pilots, who were adept at spraying liquids over crops. Floyd "Speed" Nolta ran a crop-dusting business at Willows-Glenn County Airport, eighty-six miles northwest of Sacramento, California. Before the war, he had perfected a method of dropping rice seed and fertilizer from his Jenny JN-4 biplane, and in 1952, he was flying a Boeing Stearman for his crop-dusting business.

Ely approached Nolta about his idea, and within a week, the latter had cut a hole in the bottom of his biplane and installed a 160-gallon tank with a hinged gate, a snap, and a pull. His brother Vance flew the plane over a controlled burn for a demonstration with Ely. He later documented how Vance, flying low over the fire, pulled the rope and put it out.

The first registered free-fall drop on a forest fire occurred on August 12, 1955. A crew was fighting the Mendenhall fire on the west side of Bald

Available postwar in large numbers, the rugged Boeing Stearman training aircraft were perfect for firebombing.

Mountain when they heard a plane and looked up in time to see a flood of water come down upon them and the edge of the fire. The plane that completed the drop was Boeing Stearman N75081, the first registered free-fall airtanker in the history of U.S. aviation.

Nolta, working with his brother, undertook further water drops in the Mendocino region throughout the 1955 season. Their success led Ely to form the Mendocino Air Tanker Squad. Based at the Willows-Glenn County Airport, the squad employed local agriculture pilots flying Boeing Stearman and N3N biplanes. In 1981, on the twenty-fifth anniversary of the founding of the squad, the Forest Service built a monument to Joe Ely and the pilots at Willows Airport.

Finally convinced of the value of aerial fire suppression, in October 1956, the U.S. Forest Service contracted four Boeing Stearmans and Grumman Avengers, each fitted with a specially designed release valve in the bottom of their holding tanks, to cascade both water and chemical retardants on wildfires.

The Pacific Northwest suffered a record wildfire season in 1958. With 3,305 square miles (85,591 hectares) scorched, it would be the worst in

British Columbia's history — until 2017. Unlike the province of Ontario, where government ownership fostered a direct role in the development of aerial tankers by the OPAS, in British Columbia, it was commercial air services that led the way. Six waterbombing DHC Beavers were contracted from PWA, and helicopters from Oregon — two- and three-place Bell and Hillier models — dropped water-filled paper bags on lightning strikes but with little success.

Art Seller, the owner of Skyway Air Services, a general aviation company in Langley, British Columbia, had been so stimulated by the ferocious 1958 fire season that he went to California to look at aerial firefighting. He returned home to his crop-and-budworm spraying business and declared it was too dangerous for them to ever do.

In a year, that was exactly what Skyway Air Services was doing. An infestation of spruce budworm on Vancouver Island led Seller to lease three Grumman Avenger TBMs. The aircraft impressed him enough that in 1957, when Crown Assets Disposal put the Royal Canadian Navy Avengers up for sale, Seller took a chance and bought seventeen. The Fairey Aviation Company in Victoria fitted the torpedo bombers with six-hundred-gallon

In 1957, when the Royal Canadian Navy Avengers were put up for sale, Art Seller bought seventeen and had them fitted with six-hundred-gallon belly tanks.

belly tanks and converted four Boeing Stearman crop dusters to drop water or chemical suppressants as well. Art liked to say that the firm's fleet of twelve Avengers and six birddog planes was the world's largest bomber fleet of its kind.

Tom Wilson, who was part of Art Seller's Skyway crop-dusting crew, demonstrated water test drops on August 5, 1958, for the British Columbia Forest Service (BCFS) at Langley. They approved, and the next day he flew to Bear Creek, Harrison Lake, for the first day of firebombing. Drop technology then, Wilson remembered, was basic. One filled up the Stearman with water and flew out to make the drop from a single door. The problem was that the pilot had little flexibility in making the drop.

Wilson recalled,

> The first fire was on the east side of Harrison Lake. We used the Pretty Timber airstrip for loading water and stayed in their bunkhouse, enjoying wonderful food at the cook house. The Boeing Stearman's tank was in the forward cockpit, which was stripped of the controls, instrument panel, seat, etc., to install the hopper. There was no centre-of-gravity problem. You sort of dive bombed, holding the speed around ninety miles per hour, concentrating on the target that one would select as we had no birddogs in those days. From August 6 to 24, 1958, I flew 103 hours, fifty-five minutes in eighteen days. The big day was August 17 when I flew twelve hours, thirty minutes. My base pay was $250 per month, flight pay $10 per hour.[3]

In what must have been manna dropped from heaven for Art Sellar, the BCFS awarded Skyway a $200,000 firebombing contract for the 1963 fire season. He marshalled twelve Avengers and four birddogs plus crews and radio-control trucks for it. Three Avengers with five-hundred-gallon capacity, one birddog aircraft, and a radio truck formed a base unit, each based at Smithers, Prince George, Kamloops, and Cranbrook. The Forest Service maintained a ten-thousand-gallon Bentonite mixing pit at Prince George for the Avengers and positioned smaller plywood tanks at other interior airports. Tom flew the Avenger CF-KCL; his targets were Bowen Island and Port Mellon.

Having outgrown the Langley airfield that served as its base, Skyway moved to Abbotsford Airport, southwest of Vancouver, in 1965. After suffering a stroke, Art Seller sold the airtanker side of the business in 1969, keeping the flight school with his son Dave. He kept a Boeing Stearman as his personal plane after retirement, flying it up until his death in 1998. David Seller donated the aircraft to the Canadian Museum of Flight, Langley, at a signing ceremony on November 3, 2016.[4]

Martin Flying Boats

Easily the largest wartime aircraft orphaned by aviation technology were the giant Martin flying boats. In 1938, when the U.S. Navy wanted a "flying dreadnaught" — a heavily armed flying boat — it was an opportunity for Glenn Luther Martin to demonstrate the value of his flying boats. By the time five were launched in 1944, land-based aircraft had made flying boats obsolete, and the Navy used them as troop and freight carriers. In line with the routes they served, the aircraft were given "Pacific" names. There were five of these — the *Philippine Mars*, the *Caroline Mars*, the *Marshall Mars*, the *Marianas Mars*, and the *Hawaii Mars*.

The *Marshall Mars* was lost near Hawaii on April 5, 1950; the four remaining aircraft flew until the summer of 1956, when they were beached at Naval Air Station Alameda on San Francisco Bay, waiting to be turned into scrap metal.

Before he became the chief pilot for MacMillan Bloedel, the dominant company in the British Columbia forestry industry, Dan McIvor had been flying for the BCFS, transporting firefighting crews and supplies to fire sites. In 1959, he saw that the water load dropped by the Beavers was incapable of penetrating the thick carpet of forests in the province beneath which fires raged unhindered. What was needed was a deluge so powerful, it could break through the forest's armour to douse the fire beneath. McIvor had an epiphany: the four Martin JRM-1 Mars flying

Orphaned after the war, the giant JRM-1 Mars became the world's largest operational water bomber.

boats beached at the Alameda Naval Station could be adapted to scoop up and drop water on wildfires.

MacMillan Bloedel initially turned down McIvor's recommendation that the aircraft be purchased. Former RCAF Air Vice Marshal Leigh Stevenson had been a forest ranger in 1921. Now on the board of MacMillan Bloedel, he declared that the Martin Mars were ideal for waterbombing. This convinced the company to finance the venture, and in July 1959, the consortium Forest Industries Flying Tankers (FIFT) was created to buy and operate the four Mars waterbombers. The first very large airtankers (VLATs), all four aircraft were purchased for US$100,000 — a bargain considering they had originally cost the U.S. Navy $3.5 million each.

U.S. Navy crews flew the disused flying boats to the Patricia Bay seaplane base near Victoria, where Fairey Aviation undertook their conversion to waterbombing. A six-thousand-gallon plywood tank was installed in their cargo bays, and retractable pick-up scoops were added on either side of the keel to allow uploading of water while the aircraft was taxiing. Two dumping hatches were located in the nine-foot-square freight doors in the sides of the aircraft. With a length of 117 feet, 3 inches (thirty-five meters), a wingspan of two hundred feet (sixty meters), a width of thirteen feet six inches (six meters), and height of thirty-eight feet, five inches (twelve meters) afloat and forty-eight feet (14.63 meters) beached, the majestic Mars aircraft was the world's largest operational flying boat and waterbomber. Powered by four 2,500-horsepower Wright Duplex Cyclone eighteen-cylinder engines, it had a cruising speed of 160 miles per hour, and its water load capacity equalled that of eighty-eight Beavers or ten Grumman Avengers. Able to be airborne within twenty minutes of a call and operate at 1.5 minutes per mile (one minute per kilometre) of distance between water source and drop zone, Glenn Martin's monstrosities were well suited to their new roles.

After a series of flight tests in the spring of 1960, FIFT stationed the first Mars at its Sproat Lake base on Vancouver Island. That summer, it responded to six fires, dropping 127,000 gallons of water, despite some mechanical problems. At the season's end, W.B. Gayle reported to the MacMillan Bloedel board that the Mars, for its size, was stable, seaworthy, and manoeuvrable. Glenn Martin would have felt vindicated.

Unfortunately, while fighting a fire, the *Marianas Mars* crashed into Mount Moriarty near Nanaimo on June 23, 1961, and the following year, the *Caroline Mars* was damaged beyond repair in a storm. But the *Hawaii* and *Philippine Mars* entered service in 1963 and doggedly fought wildfires for more than forty years.

MacMillan Bloedel was acquired by the American giant Weyerhaeuser in 1999, and when the softwood lumber market collapsed in 2006, decimating the homebuilding industry, the two Mars bombers were put up for sale. They were acquired by the Coulson Group in 2007, with CEO Wayne Coulson vowing to transform the Mars operation into a business that could battle wildfires worldwide. His company invested heavily in maintenance, fuelling, and support equipment for the aircraft, upgrading the *Hawaii Mars* with a glass cockpit so it could safely operate in high-density U.S. airspace. Water drops were made on a calibrated grid so the aircraft could receive U.S. Air Tanker Board (ATB) certification.

Palaeontologists say that coexisting with the dinosaurs were furry little mammals that survived the asteroid impact that wiped out most of the life on Earth sixty-five million years ago because they were adaptable to climate change. So too did the Martin Mars give way to the nimble single-engine airtankers (SEATs), which boasted quicker turnaround times and were able to make shorter scooping runs on smaller bodies of water. The *Philippine Mars* was retired in 2012 and will be preserved at Pima Air & Space Museum in Tucson, Arizona.

The *Hawaii Mars* had its last fire season in 2015, and after a donation of $250,000 from the provincial government on March 28, 2024, it was announced that the aircraft was to be placed in the British Columbia Aviation Museum in Victoria. On August 11, 2024, Harbour Air pilot Pete Killin and alumni Rick Matthews took the waterbomber on its last flight from Port Alberni to the Saanich Inlet, Victoria. It was flown to Campbell River and Powell River, then to Comox and down the coast. According to the British Columbia Aviation Museum, its flight route was significant. The *Hawaii Mars* passed over several B.C. communities to commemorate the forest industry's original establishment of the Martin Mars water bomber program to fight forest fires.

In the final portion of the flight around Victoria, the *Hawaii Mars* was joined by the Canadian Snowbirds from 431 Squadron, its nine CT-114 Tutors flying in formation with the water bomber. Crowds lined various locations along the route, their number estimated to be over ten thousand, to witness the final arrival of the Martin Mars aircraft. It touched down for the final time at 7:00 p.m. PDT to the roar of the appreciative crowd that lined the foreshore and gathered in hundreds of boats in the bay, waiting to witness the arrival of the seventy-nine-year-old aircraft that had served as an iconic firebomber. "It's kind of sad that it's the end of the story," said Killin, "It's a new chapter coming [for the aircraft] … it's going to be good, people will get to see it," he added.[5] Would that Glenn Martin and Dan McIvor were there too.[6]

Retardants

The early airtanker pilots, most of them crop dusters, had to work out the techniques to successfully deliver their loads. Drops over forest fires required passes deep in canyons or toward high ridges. On hot days, even dropped at fifty feet, much of that water evaporated before it hit the ground, so the pilots liked to fly at five to ten feet above the trees.

There were the ever-changing treetop heights, smoke-filled air, and treacherous convection currents. The loaded weight of a Stearman was 2,950 pounds (133 kilograms). Dropping the water load of 125 gallons caused an immediate 50 percent change in weight, allowing the pilot to use the rapid rise to clear any approaching ridge or tree line. The problem was that the former training aircraft had not been constructed to safely recover from dives and violent low-level manoeuvres. Without radio, the pilot didn't know where the ground crews were, and with the early line-of-sight radio transmission stations blocked by the mountains, neither did the dispatcher back at the base.

The service wanted to maintain a safety margin of seventy-five feet above the trees but not lose the effectiveness of the water. The solution was to mix water with sodium calcium borate, producing a compound that made it to the ground with a melting point twice as high as the nine-hundred-degree ignition point of a forest fire. As the white material stuck to brush and reflected heat, it was employed on the flanks of fires to control spread.

Marketed as "Firebrake," sodium calcium borate was one of the first chemical retardants to be dropped on wildfires in the early 1960s. It was only utilized for a short time because, although it was white when it dropped, it quickly blended in with the vegetation, so no one could see where it had been deployed. As well, its high cost, the fact that it was a soil sterilant, and its abrasiveness to equipment made it unpopular. Although use of the chemical was abandoned, the nickname "Borate Bombers" for the planes dropping retardants stuck.

Bentonite, another thickening agent, became the most widely used short-term retardant in British Columbia. The BCFS built stations at airports in the province's interior for mixing the retardants and loading tankers, and by the end of the decade, more efficient long-term ones like Phos-Chek and

Fire-Trol were coming into use. These salt-based materials gave off ammonium gas when heated, robbing the fire of oxygen and slowing combustion even after the water itself had dried.

The retardant slurry is typically dropped along the fire's edge to create a barrier that cools the edges and allows ground crews to get closer and stop the fire's progress. It's usually coloured red or high-visibility pink to show where the retardant is in the trees and allows for other pilots to "tag" on to extend or turn the line. Contemporary retardant properties can last for many hours to help firefighters. Water is considered more of a "short-term" product and is typically dropped directly onto the flames as a suppressant by the water-scooping airtankers. Generally, more than one "scooper" aircraft is dispatched to a fire, and they can be extremely effective in suppressing a fire, especially when a water pickup source like a lake or river is nearby.

CHAPTER 8

Birddogging

EVEN BEFORE THE D-DAY LANDINGS, THE U.S. ARMY WAS SEARCHING FOR A light aircraft that could fly low and slow over enemy formations and report on artillery fire targets, as the Taylorcraft Auster did for the British and Canadians. The Cessna 305A was chosen, and in the competition to name the aircraft, photographer Jack A. Swayze's winning entry was *Birddog*. The U.S. Forest Service bought the C305A (L-19) for fire patrols and coordinating aerial firefighting.

Until 1960, firebombing success, as Tom Wilson knew, relied on a pilot's foolhardiness or direction from the ground — if that was available. When the Grumman torpedo bombers were put into service as airtankers in the late 1950s, a staggering death toll — twenty-two out of twenty-eight pilots in the first year — convinced the Service that a lead plane was essential. The birddog or air attack position was created as the airborne command and control of a fire response team, evaluating the situation in the air and on the ground and directing assets to the fire's hot spots.

Mary Barr was hired to be the Forest Service's first birddog pilot. She began her aviation career chasing moose off the runway as a little girl. She took almost any opportunity to get herself off the ground — washing dishes, working as an aircraft maintenance engineer (AME), giving flying lessons, and even transporting prisoners. She began by spotting fires on her own and phoning them into the Forest Service.

In their hopelessly inadequate Cessnas, the first birddog pilots were blinded by smoke or by ash raining down. Shifting winds and sudden updrafts could roll their small plane, flip it upside down, or slam it to the ground. Barr knew that as a female pilot she had to be twice as good, twice as calm in the face of upsets, to get anywhere. The job, said a pilot, is like playing chess — except the chess board below you is on fire. She was proud that the other pilots — all male — would follow her into this scenario from hell with complete trust. Mary Barr's death from complications of dementia, on March 1, 2010, at eighty-four, went unreported — until now.

The search for a satisfactory birddog aircraft in 1973 led California (CAL FIRE) to acquire twenty Cessna O-2 aircraft that had been used during the Vietnam War. That year, a birddog type that for the first time matched the speeds of the A-26 and DC-6 tankers was introduced: the Ted Smith Aerostar 600. Extremely light and manoeuvrable, with good acceleration characteristics that made it both safe and fun to fly, it was a perfect fit for the role. The aft, midwing design permitted excellent visibility forward, downward, and to the side, allowing for a commanding view of the fire area. The Aerostar was able to safely operate at the same lower air speeds that the airtankers flew during their final approach to the fire.

Rockwell International updated it as the Turbo Commander, with its exceptional visibility for the air attack pilot and air attack officer (AAO), their seats situated well ahead of the engines and propellers. In 2024, the Turbo Commander TC-690A, pressurized and certified to a ceiling of 31,000 feet, was the preferred birddog of airtanker fleets. Mary Barr would have loved it.

Art Kirk was British Columbia's and Canada's first birddog officer. The first Canadian birddog school was held at the Green Timbers training facility at Surrey, B.C., to prepare for the 1960 contract season. The week-long course was supervised by Art and involved forest protection officers as well as pilots from Skyway. Flight exercises were conducted from Skyway's base at Langley.

Ray Horton: A Birddogger's View

Working for a charter company in Fort Simpson, Northwest Territories, in the summers, Ray Horton flew supplies to the firefighting base camps. There he became intrigued by the operations of the Conair planes and their firefighting pilots.

He was hired by Conair to fly birddogs and over the next eighteen years worked as a line pilot, gaining experience on a variety of aircraft, including both birddogs and airtankers, eventually becoming chief pilot. Horton described the birddog role to me thus:

> In those days, new pilots would typically start on the birddog plane to learn the ropes of firefighting or on the DC-6 as a first officer. The birddog plane is much lighter and more manoeuvrable than the heavier airtankers and works as a scout or reconnaissance aircraft to check out all the potential drop areas on a fire. I was fortunate to fly the Aerostar in my first year, which is as wonderful to fly as it is to look at. At the time, it was the fastest in-production general aviation twin engine airplane, and it suited its role as a birddog perfectly. It was faster than the Firecat and not much slower than the behemoth DC-6.
>
> There are typically two crew members in a birddog: the pilot and the Ministry of Forest representative — the air attack officer (AAO). The AAO is a highly trained and knowledgeable ground firefighter who specializes in fire behaviour and control techniques. Among the many things an AAO considers while fighting a fire are fuel (timber) types, moisture code indexes, temperature, wind, relative humidity, forecast weather, terrain, and slope exposure. They consider all these variables while forming their plan of attack on how best to fight a fire. As the government representative, they provide the on-scene oversight of the cost of fighting a fire with airtankers. When responded to promptly, most fires can be effectively fought using

fixed-wing aerial firefighting tankers. However, many of the variables outlined sometimes stack up against the efficient and/or safe use of airtankers, and it's the birddog crew's role to assess that and take appropriate action — which may even be no action at all.

To be effective, water or retardant drops from aircraft need to be performed within a certain range of parameters. As a rule, and depending on aircraft type, pilots strive to make retardant drops at an altitude of between 125 to 200 feet above the tree canopy and at between 110 to 140 knots of airspeed. Outside factors such as wind, terrain, visibility, exit routes, and aircraft performance must all be taken into consideration when planning an airtanker drop on a fire. The birddog pilot will fly at the exact same speeds and altitudes as the airtankers while executing and assessing their reconnaissance runs.

The birddog pilot and AAO work as a team to provide a safe plan of attack for fighting a wildfire with fixed-wing aircraft. It is their job to ensure they can safely bring a heavily laden airtanker overhead and then down low onto the fire for a drop and most importantly guarantee the safe exit back up to altitude again. The birddog must check out all runs for safe approach and exit patterns ahead of the airtanker, all the while assuming the airtanker could lose an engine and/or be unable to drop their load, yet still be able to clear terrain while exiting the area. Without question, the birddog's role is the most important piece of the firefighting team. They must also act as air traffic controllers and "stack" airtankers into holding patterns overhead the fire when multiple aircraft are dispatched at the same time.

The birddog aircraft and crew typically spend an entire fuel cycle at a maximum altitude of one thousand feet in an orbit overhead the fire and demonstrate "show-me" or "dummy runs" of the expected drop profile to the airtankers overhead. This low-level, demanding work requires a high level of vigilance by the crew the whole time they operate at the fire. Needless to say, it can be very fatiguing on long days. This is why there are extremely strict standard operating procedures (SOPs) that all airtanker crews must learn and adhere to, to keep the operation safe.

When performing dummy runs for an overhead airtanker, the birddog must demonstrate good timing and an awareness of the airtanker's position. The airtanker, being much heavier and less manoeuvrable, is limited in performance and agility when the pilot attempts to get in a position to observe all aspects of the dummy run conducted by the birddog. Therefore, the birddog pilot must ensure he observes the airtanker is in a good position to watch an impending run. If the tanker is out of position or is subject to parallax (offset) vision, the dummy run will be wasted and will have to be performed again.

A proper dummy run takes three to five minutes to perform, depending on terrain and visibility in the smoke. Active wildfires can move a fair distance in a brief period, so while safety is paramount, efficiency of time management is also important for a successful result. No question, a good birddog crew can make an average airtanker pilot look good!

The dummy run is also a signal to ground firefighters that an impending run is about to occur. Birddog planes are equipped with a siren, which is used by the crew during the low-level dummy run to warn the ground crews that a drop is pending. Airtanker pilots will never intentionally drop on people or ground crew. While the retardant or water drop may not harm the crew, tree branches that may be broken off by the load could well impale a person.

Well-flown circuits, using physical features around the fire (lakes, ponds, cut lines, tree types) to describe where to turn and line up on a fire, are all important birddog techniques. "Maximum amount of information, minimum number of words" is key when describing runs and providing good situational awareness to airtanker pilots. Any direct safety threats are also pointed out by the birddog crew during the dummy runs, such as nearby power lines, high terrain, or unusually tall or dead trees (snags) in the drop area — all of which can be difficult to see when sunlight glare and/or smoke is an issue. They must never assume the airtanker pilot will see something the birddog crew think is "obvious."

The birddog spends three to four hours over a fire at a time, but the airtanker is only there for a few minutes before being sent back for another load or to another fire. Therefore, what might become or seem "obvious" to the birddog crew may not even be seen by the airtanker pilot(s) unless specifically pointed out as they strain to watch and understand the dummy run.

Once the airtanker pilot understands the run and begins to set up a circuit for the drop, the birddog clears the tanker down into the birddog's airspace and sets up a loose formation to follow the tanker in on the run. This is another safety net for the airtanker that allows the birddog crew to keep their eye on the pattern the tanker is flying and be able to immediately correct any possible mistakes, such as altitude, airspeed, or the incorrect run direction. In addition, all drops are recorded and rated for accuracy by the AAO on a drop sheet. Aerial firefighting, as effective as it can be, is expensive, as is the retardant, so the airtanker team work hard to ensure drop accuracy.

Like an orchestra conductor, the AAO chooses which attack to use. The three basic methods in the attack are direct attack, parallel attack, and indirect attack. Scoopers make high-volume drops directly on the fire or its edge in direct attack. With parallel attack, drops are made a short distance from the fire's edge, building a fire line from an active fire front and parallel to it. In indirect attack, drops are made away from very large fires, utilizing any natural barriers.

As the target is approached, the AAO begins the countdown "3, 2, 1, bombs away NOW" and the airtanker pilot triggers the load on hearing the "NOW." Hopefully, the assessment is a "bull's eye" for the airtanker pilot. By working as a formation, the aircrew of each individual aircraft adjusts the placement of their water drop with reference to the drop from the aircraft they are following.

With fire, smoke, and otherworldly creatures swooping down, if the whole exercise were put to music, Richard Wagner's "Ride of the Valkyries" would suit.[1]

Conair Birddog Turbo Commander. The Birddog is the airborne command and control centre of a fire response team, evaluating the situation and directing assets to the fire's hot spots.

As with most facets of aviation, techniques and standard operating procedures for aerial wildfire suppression took many years and even decades to develop. Sadly, many of the dos and don'ts came about from serious mistakes, misjudgements, or miscommunication that resulted in much loss of life throughout the history of aerial firefighting. After-action debriefs are now mandatory and are used to correct any errors or unanticipated events that may have occurred during a fire action. Input is received from the birddog and airtanker crews, as well as ground crews, when necessary, to ensure safety is ensured and no unsafe action, inadvertent or otherwise, is repeated. This is where lessons learned are discussed and often put into future practice.

Wildfires are typically fought by many aircraft types with various roles, all operating within the relatively small airspace surrounding a fire. Helicopters, birddogs, airtankers, and even the odd and highly illegal stray aircraft (and, increasingly, drones) that wander in "to have a looky-loo." This can lead to numerous near misses if all rules of engagement and standard operating procedures are not followed precisely by those involved.

Midair collisions have occurred far too frequently over the years, with a resulting loss of life. One near miss Ray Horton encountered was when the airtanker pilot he was formatting on mistakenly turned in the wrong direction during his exit, which put him nose-to-nose with Horton in the birddog. "With both of us below two hundred feet above ground, it left precious little room for me to duck between him and the treetops to avoid a collision. Needless to say, the after-action debrief was certainly full of heated debate and learning following this incident." Nowadays, traffic collision avoidance systems (TCAS) equip all airtanker aircraft, greatly enhancing traffic awareness.

As can be imagined, smoke can also cause a serious hazard for aerial firefighting aircraft. During bad fire seasons, hundreds of square miles often become blanketed with thick smoke. Ray recalled one memorable occasion in 1981, in the days well before GPS, when a great number of fires were burning throughout the Fort Smith, Hay River, Fort Simpson, and Yellowknife areas of the Northwest Territories:

During a repositioning flight from Hay River, where visibility was about ten miles, to Fort Smith, we found the visibility began to worsen the closer we got to Fort Smith. A DC-6 was following about ten minutes behind me, and as forward visibility continued to drop, we elected to navigate toward the Slave River, which leads to Fort Smith and its airport on the edge of town. From our previous flights to Fort Smith, we were both aware of a seismic line that happened to run from the river's bank directly to the threshold of the runway.[2] The plan was to use these ground reference points to help with our navigation.

At about twenty-five miles north of Fort Smith, day turned into night. The blanket of smoke in the area became so thick that no sunlight whatsoever was able to penetrate, and we descended into complete darkness. I turned on all the aircraft instrument lights and warned the DC-6 of the impending conditions ahead. Those pilots told me later they thought I was exaggerating and so were caught totally off-guard themselves as they scrambled to find the various instrument lights in the pitch-black cockpit.

The huge fire that was causing the associated darkness was so massive, it was also creating its own weather system, and light drizzle was beginning to fall. This was something I had never witnessed before, so I failed to comprehend how it impacted the forward windscreen — it became covered in ash from flying in the thick smoke. Although dark, we managed to continue navigating from our side windows along the river's edge to the seismic line leading to the runway. I could barely make out the runway lights, which had been turned to maximum brightness. That was when I realized the ash had been mushed by the falling drizzle and fully opaqued the forward windscreen, reducing visibility to almost nil. Side-slipping the airplane allowed me to look out the side cockpit window and down to a safe landing.

> Watching the behemoth DC-6 do the same a few minutes later was a sight to behold.

Whether flying a Turbo Commander, DC-6, Martin Mars, or (in the next chapter) a Canso, aerial firefighting pilots will warn you that their profession is not for everyone. Home in the summer fire season is usually a cheap motel with a kitchenette or an RV/motorhome, far from your family for anywhere from 90 to 120 days. The running joke in the industry is as soon as you schedule your spouse to visit, you get a fire call hundreds of miles away! In the winter, the pilots are grateful for contract positions flying somewhere else.

So, why would anyone choose to be an aerial firefighter? When I asked him, Brad Blois spoke for all in his profession: "It is to be a part of something bigger than yourself, helping protect people and communities. I can look back on what I have done and feel a great deal of pride and satisfaction in my work."

CHAPTER 9

Catalinas and Cansos

Catalina

Some aircraft achieve greatness, like the Spitfire and Mustang, and others have greatness thrust upon them — like the Catalina. Throughout the Second World War, the flying boat was involved in actions that changed the course of history. A Royal Air Force Catalina crew located the German battleship *Bismarck*, and a U.S. Navy Catalina crew spotted the Japanese task force approaching Midway. Squadron Leader Leonard Birchall, an RCAF Catalina pilot, sighted the Japanese invasion force making for Ceylon (Sri Lanka), and another RCAF Catalina pilot, Flight Lieutenant D.E. Hornell, was posthumously awarded the Victoria Cross in July 1944.

The Catalina, a twin-engine, parasol-winged flying boat, had been designed for a United States Navy competition in 1933 by Isaac Machlin Laddon, Consolidated Aircraft Corporation's aeronautical engineer. The Navy wanted a military patrol flying boat with an operational range of three thousand miles (4,828 km) and a cruise speed of one hundred miles (160.9 km) per hour. Consolidated submitted Laddon's design, the Navy placed an order for sixty aircraft, to be known as the PBY-1, and the company began to plan production at its new facility at Lindbergh Field, San Diego. The

PBY designation (P for Patrol, B for Bomber, and Y for Consolidated) recognized the aircraft's ability to operate not only in the patrol role but also as a bomber. The story is probably apocryphal, but after seeing Santa Catalina Island near San Diego, the British Direct Purchase Commission named the flying boat "Catalina."

Canso

In early 1941, Canadian Vickers Ltd., in the Montreal suburb of Cartierville, and Boeing Aircraft of Canada, in Vancouver, began production of the PBY-5 and PBY-5A, named "Canso" after the Nova Scotia strait. Committed to shipbuilding, Canadian Vickers left the aircraft business, and on October 3, 1944, its Montreal plant was incorporated as Canadair under federal government management. When production ceased on May 19, 1945, Canadian Vickers/Canadair had built a commendable 369 Cansos, and the Boeing plant in Vancouver 240 Catalinas. The last Canso was retired from RCAF service on November 29, 1962, and immediately put into fire suppression service with the Government of Quebec as CF-PQO.

The Canso flying boat really came into its second life as a water bomber on May 19, 1960, when Austin Airways of Timmins, Ontario, collaborated with Montreal's Aircraft Industries of Canada to mount 350-gallon underwing tanks on Canso CF-JTL. Filled through four-inch tubes, the "short term retardant" (i.e., water) exited from tank tail caps.

That same year, when he converted a PBY to firefighting duties, Field Aviation's J. Knox Hawkshaw invented the Canso water bomber scooper and patented it in his name.[1] Liston Aircraft of Klamath Falls, Oregon, built the first Catalina with internal tanks in 1962, and Field Aviation and Fairey Aviation of Canada began converting PBYs for the firebombing role for Canadian provinces.

The Cansos modified by Field Aviation incorporated an eight-hundred-Imperial-gallon interior bomb tank filled from a single hull-mounted hydraulic probe. The probe was activated from the cockpit once the Canso was on "the step" on the water. It would fill the bomb tank in twelve to fifteen seconds. The pilot then had the option to drop the water load in one

Until the advent of the CL-215, Cansos were the workhorse firefighting aircraft in Canada.

of several combinations using the two flush hull-mounted bomb doors. The normal drop was a "salvo," where both bomb doors opened simultaneously, dropping the entire load.

Alternatively, a pilot could drop a "single door" on a small spot fire or do a "string drop," releasing the two doors one after the other to extend the length of the water drop.

In a country with an abundance of lakes and where budgetary constraints prohibited purchase of chemical retardants, until the Canadair CL-215, Cansos were the workhorse firefighting aircraft. Able to scoop water from a nearby lake or carry fire-retardants from airstrips, the old patrol aircraft combined a large water capacity with flexibility. The OPAS evaluated one of the first Field Aviation conversions, but, wedded to the Beavers, used it only for transportation. New Brunswick chartered two Cansos and borrowed two others from Quebec during the severe 1965 fire season.

By 1968, thanks in part to the increasing use of aircraft, most of Canada's 620 million acres of productive forestland were afforded some level of protection. The average fire burned an area of 164 acres (sixty-six hectares) between 1962 and 1966, down from the 1943 to 1947 average of 220 acres (eighty-nine hectares) and almost one-third less destructive than the 1923 to 1927 average of 470 acres (190 hectares).

Tim Garrish

Tim Garrish, who retired in 2023 with over fifty years of experience of bush flying in winters and aerial firefighting in summers, with an accident-free total flying time of over sixteen thousand hours, knows the Canso very well.

Graduating from Simon Fraser University in December 1976, Tim wondered what to do next. He thought his degree in economics and geography had limited application, other than for reading the newspapers with a critical eye.

In the spring of 1977, he got a call from Pete Cowie in Fort Simpson, Northwest Territories. Pete had a Cessna 185 CF-TFO aircraft on amphibious floats. Wolverine Air Ltd. (WAL) had a contract for a C-185 amphibious aircraft with Northwest Lands and Forests (NWLF) for a birddog aircraft on firebombing operations to be based in Yellowknife. But the pilot who Cowie had lined up had failed to show. He called Garrish on the off chance he was available.

Tim arrived in Yellowknife the next day in time to ride along in the backseat of CF-TFO while Pete worked a fire. "I closely observed the operation. Pete left Yellowknife the next morning, and I was in the firefighting business, where I would be for forty-four seasons."

Tim recounted:

> The Canso was given another lease on life after the war ended. It was adapted first for charter/scheduled services serving communities in remote areas along the coast of British Columbia and around Hudson Bay. It got another lease on life as a water bomber on forest fires in Canada for many years.
>
> The Canso had a range of up to twenty hours with full fuel when utilized on long-range patrols during the war. However, in a firebombing role, the normal fuel load was

four to five hours to supply the reliable Pratt & Whitney 1830 radial engines rated at 1,200 horsepower. The gross weight in civilian use was 30,500 pounds (13,834 kilos); however, the military operated the aircraft to higher weights. Single-engine performance at the normal gross weight was marginal. At the higher gross weights, it would have been nonexistent. One major benefit for the fire-bombing pilot was that in the event of an engine failure, one's first action was to drop the water load and significantly reduce the aircraft's weight.

Avalon Aviation based in Red Deer, Alberta, was a major Canso operator until 1980, when the operation was sold and moved to Parry Sound, Ontario. Avalon had summer contracts in Ontario, the Northwest Territories, and Alberta, along with one aircraft operating in Norway each summer. Each winter, one Canso was ferried down to Chile for a contract there as well. The extended fuel range came in handy on the ferry flights to Norway and Chile.

The Avalon Aviation Cansos also had a compressed air system on board to inject a dry powder called Ten-O-Gum into the water load prior to dropping. This pink-coloured water provided the pilot with some reference as to where the load had gone on the ground for reference for subsequent drops.

A major weakness of the Canso in rough water was the nose gear doors. There have been several failures of the nose doors over the years, with dire results. Hence, a Canso pilot was very aware in rough water operations to maintain a nose-up attitude to keep the nose doors out of the water or a nose-down attitude in higher wind/water conditions so that the hull, ahead of the doors, took the brunt of the rough water.

Not having nose wheel steering, differential power was used for taxiing the Canso on the ground as well as docking on the water. On occasion, the Canso would be used to tanker fuel to remote communities in Ontario. One tank would have aviation fuel for the engines, while the other tank was isolated and used to haul diesel or furnace oil. The challenge was docking the aircraft and then filling up forty-five drums by gravity from the aircraft tank.

Normal cruise speed was a modest 105 knots. When scooping water, the aircraft would rotate off the water at a minimum speed of sixty-five knots. Attempting to get the aircraft airborne below that speed would inevitably result in the aircraft entering a "porpoise,"[2] which could be catastrophic. Several Cansos have been lost due to porpoises over the years in commercial operations. Avalon had one severely damaged in a training incident in May 1978. A new captain was getting his final assessment ride by Bob Murdoch at Sylvan Lake, Alberta. It didn't go well.

Unfortunately, in Tanker 1, CF-CRR, the airspeed indicators were graduated in miles per hour and not knots per hour as the rest of the fleet. Bob must have forgotten this, because when he saw sixty-five miles per hour on the airspeed, he started yelling for the new captain to "pull her off" and get flying, which he then attempted to do. Sadly, sixty-five miles per hour is only some sixty knots, and the fully loaded Canso was not ready to fly. It tried. When the pilot presented the Canso wing to the airflow, something had to happen. The aircraft went into an immediate porpoise.

I was overhead on a training flight as well and observed the excitement as the Canso disappeared into a cloud of spray as it porpoised down the lake. When it stopped, it had water-looped and was facing in the opposite direction with an obviously broken spar on the right wing. The aircraft was immediately beached because water was flooding in through the blown-out cockpit floor inspection window, amongst other places. The aircraft was refloated off the beach the next day, a Sunday, and hauled up on shore at the community park at south end of Sylvan Lake.

The aircraft required a fair amount of pilot input on the controls to fly it. It had no hydraulic assist and was especially slow and heavy on aileron response. A major input of aileron would be required to raise a dropped wing and took several seconds. Not having flaps, our profiles were fairly straightforward.

I was doing my first Canso ground school in Red Deer at the time when we received word of the accident in May 1978. Likely, it was totally preventable. The crew had arrived on base for the season to be greeted with fires happening that day. Rather than unload the spares, gear for the summer (even a motorcycle remained onboard), they went straight to work. Later, they stopped to dip their tanks on a lake (no gauges in the Canso that worked). They were low on fuel, but the captain kept going, and they figured one engine quit of fuel starvation at the drop. They lost control and went in, with both crew killed. Fatigue was no doubt a factor. The young copilot was on his first day of bombing. A letter later appeared in one of the papers from his mother basically asking what the hell was going on with this operation.[3]

The Avalon aircraft did not have functioning fuel gauges or gear position indicators. Fuelling was done with dipsticks to confirm contents. As for the landing gear, after takeoff, one pilot would go back in the cabin to confirm the main gear was indeed retracted and on the up locks. The nose gear being up with nose doors closed could be confirmed via the small inspection window in the cockpit floor. The reverse would happen on landing. Once again, a pilot would go aft, clamber over the bomb tank, and confirm through a small inspection window that the gear was down and locked.

As with any amphibious aircraft, landing on water without the gear fully retracted usually ends in disaster. A runway landing with the gear up often resulted in minor damage; the most damage was to the pilot's ego.

In the 1970s, it was normal procedure in Ontario, Saskatchewan, and Manitoba for Canso pilots to "lone wolf" forest fires (without a birddog). Other jurisdictions, like the Northwest Territories, required a birddog aircraft on all bombing operations to ensure separation of aircraft and helicopters over fires. This came about in the Northwest Territories after the tragic midair collision and loss of both crews on the two Cansos over the Pine Point fire on September 1, 1971. A helicopter had crashed into the fire, and the two Cansos came overhead separately to observe. Unfortunately, they were circling in opposite directions and collided.

The aircraft would be deployed into the field in late April or May and returned to Red Deer at the end of August. The Canso did not have heaters, so the spring deployment to base could be a cold one, and winter gear was essential. It was also inevitable that pilots would get oil dripped on them when walking under the radial engines.

The hours flown each summer would depend on the fire activity. When fire activity was intense, it was not uncommon for Canso pilots to fly three four-hour missions in a day. This was before the Department of Transport brought in limits on the maximum number of hours a pilot could fly, along with maximum duty day for hours at work. Eventually, the duty days were capped at fourteen hours, with eight hours of flight time being the normal daily limit.

CF-CRR Canso over Great Slave Lake, 1977. The water bomber had already served in the wartime RCAF and postwar Canadian Pacific Airlines.

My maximum number of drops in one day on the Canso was 105 on a fire east of Fort Smith in 1978. For comparison, in my twenty-five years flying the CL-215 aircraft, my maximum number of drops was a mere seventy-seven in one day. Despite the Canso having a basic intercom, it was rarely used due to the constant static in the pilot's ears. Hand signals were used for the power settings and gear deployments. It worked well.

Base changes were dependent on the fire activity, so we could be deployed across Canada. We ranged from Inuvik in the Northwest Territories, across Alberta, Saskatchewan, and Manitoba, and often into Ontario. Quebec had their own fleet of government-operated Cansos, as did the Government of Newfoundland prior to the CL-215 arriving.

One fire of note that made for an interesting day was in late April 1980, west of Edson, Alberta. It was a dry spring in Alberta, and the fires were very active prior to the "green up" of the bush. We were working a fairly aggressive fire in pine trees on a steep side hill, which required an exit of a few seconds through the smoke column after the drop. Just as I dropped, the fire boiled up in intensity in the pine forest. Pine trees loaded with pitch sent a burst of molten pine pitch up into the column as we flew through it. Instantly, the forward windows and wing leading edges were covered in a film of pitch that totally obscured our forward vision. On arrival back at Edson, the landing was made with side window open so I could get a little visual reference on the runway for the flare and landing. Maintenance spent several hours using aviation gas to remove the pine pitch from the windscreen. Then it was back to the fire.

For many years, the Canso was really the state-of-the-art water scooper for fire suppression in Canada, until the Canadair CL-215 came on the scene in greater numbers, beginning in 1986. However, cost is always a hindrance in aviation, and that was no exception with the CL-215. Bob Murdoch, the owner of Avalon Aviation, a major Canso operator, would extoll the cost advantages of the Canso over the CL-215, often claiming he could operate ten Canso aircraft for the price of only one CL-215. The capital costs of the Canso were minimal compared with that of a new CL-215.

• • •

The Catalina/Canso thus faded into retirement, the doughty old warriors destined for scrap heaps and museums. Having fought the enemy, human and holocaust, since 1939, they could proudly show their scars and say, "These wounds I had on Crispin's day."[4]

CHAPTER 10

The Canadairs

C.D. HOWE, THE MINISTER FOR RECONSTRUCTION, WANTED AIRCRAFT manufacturing to remain in Canada postwar, at any cost. He had the Hawker Siddeley Group take over Victory Aircraft in Toronto, renaming it A.V. Roe. In Montreal, Canadair had kept its workforce busy by assembling the North Star airliner for Trans Canada Airlines. Howe also convinced Electric Boat of Groton, Connecticut (later General Dynamics), to buy Canadair on January 11, 1947, for a nominal fee — a bargain $10 million.

Canadair built fighter aircraft, maritime patrol aircraft, swing-tail freighters, and, far ahead of its time, the tilt-rotor CL-44, which found no customers, either at home or abroad. Unable to compete with the United States and British aircraft giants, in the early 1960s, the company looked to emulate De Havilland Canada's success with the Beaver and Otter by building a bush plane. The CL-43 was conceived as a transport aircraft based on the design of the Canso.

Legend has it that Tom Watt, Canadair's marketing manager, mentioned to Richard Murdoch of the Quebec Government Air Service that Canadair was considering a twin-engine bush floatplane as its next project. Murdoch knew that the Quebec government was concerned about the age of their

Cansos. Spare parts were becoming difficult to obtain, and they were not going to survive many more fire seasons.

Paul Gagnon, the first director general of the Air Service, had been a bush pilot in the 1950s, flying for A. Fecteau Transport Aérien. He told Eric McConachie, the Canadair sales engineering manager, that instead of a bush plane, the company should consider a replacement for the Canso — but with a bigger water load. Its body had to be strong enough to withstand the scooping up of 1,200 gallons (4,524 litres) while performing the choreography of waterbombing.

Grappling with marketing the CL-84 tiltwing, Canadair was loath to take on another imaginative project with limited appeal. The company even had Knox Hawkshaw fly Field Aviation's Canso CF-PQM to its plant at Cartierville and demonstrate a water drop on the runway. Roger Lewis, the General Dynamics president, then bidding on the F-111 contract, was not enthused with the anachronism of a flying boat, its use confined to the summer fire season. The U.S. Forest Service was firmly entrenched in using retired military aircraft as airtankers carrying retardant instead of water bombers. As the CL-84 had shown, there wasn't a market for so unique an aircraft.

In 1962, the CL-204 design was reborn as a water bomber in the form of an amphibious flying boat. To pay for the development, construction, and certification of the two prototypes, Canadair needed thirty firm orders. Gagnon worked hard to convince the Quebec government to purchase twenty planes, and in 1965, the Canadian government agreed to finance 50 percent of the development costs. Called the "CL-215," the concept model was revealed to the public at the 1965 Paris Air Show. *Flight International* pointed out that it was the first large aircraft purposely built for waterbombing. It would not be until February 1966 that General Dynamics agreed to the CL-215 project — provided Canadair could find more customers.

The French government's Sécurité Civile wanted CL-215s, but only if the Canadian government would buy eight Dassault Falcon 10 executive jets. Protracted negotiations with the Quebec government led to a firm order of twenty CL-215s on May 19, 1966, followed a few days later by an order of ten CL-215s by the French government, later increased to fifteen. Spain bought eight dual-role firefighting and search and rescue Series II CL-215s, paying for them over eight years with shipments of Spanish wine to Quebec.[1]

The CL-215 made its first flight on October 23, 1967, the aircraft suitably registered as C-FEU-X. Its first water takeoff was on May 2, 1968, and it received FAA certification in the Restricted category on May 15. As in a cliché out of the Brothers Grimm, the ugly duckling water bomber was then sidelined by Canadair, which put all its energies and finances into building the Challenger 600 Business Jet. This bankrupted it, and in January 1976, the Canadian government nationalized the company.

What sold the Canadair water bombers to countries that did not have the many lakes that Canada did was the CL-215/415's short takeoff and landing (STOL) abilities. Although the CL-215 could scoop out of lakes shorter than two miles in length, there was always a risk in doing so. There is a story about a Spanish crew training on the CL-215 in Spain in 1995 who found this out after having come to a complete halt in a lake too short to

The CL-215 was the first aircraft in history designed exclusively as a water bomber. It became the backbone of many countries' airtanker fleets.

take off from again. The solution was to build a temporary road to the lake and then tow the plane overland to a suitable area for takeoff on wheels.[2]

Peter Marsh enjoyed forty-three years as a pilot, thirty-three of which were spent in aerial fire suppression, flying both the CL-215 and CL-415. He remembered that the constant attention and adjustments to operate the 215's radial piston engines contributed to a high crew workload. On hot days, Marsh recalled that the flight deck was a sweatbox; heavy flight controls made crosswinds a handful. Refuelling between missions required one of the crew to climb on top of the aircraft and lower a rope to pull up the refuelling nozzle. Engine oil replenishment was also a chore because oil was added from twenty-litre buckets. Oil capacity for each engine was 114 litres, depending on mission length and engine condition, and each engine could require anywhere from 4.5 to 20 litres. To Marsh, it all just made a long day longer. But, he conceded, the aircraft fulfilled its mission very well.

So thought Yugoslav Army pilot Major Radovan Katanic, who, on July 14, 1983, flying a CL-215, made 225 drops on a wildfire, totalling 317,760 gallons, in a single day, a record still unbroken.

The first aircraft in history designed exclusively as a water bomber, the CL-215 was an all-metal, armoured boat with wings. Critical parts were made with treated aluminum, allowing it to operate in fresh and salt water. Its Pratt & Whitney R-28000 eighteen-cylinder radial engines, first built in 1939, were reliable. Able to carry a maximum 1,412 gallons (6,419 litres) in the two fuselage tanks below the wing, it could be ground-filled in two minutes or scoop-filled in ten seconds while planing at seventy knots.

With a gross weight of 43,500 pounds (19,731 kilos), it was a vast improvement over the Canso, both in the water load carried and cruise speed. It had a 50 percent higher cruise speed of 150 knots, as opposed to the 105 knots of the PBY. It could scoop 1,200 Imperial gallons (5,455 litres) versus the Canso's 800 Imperial gallons (3,636 litres) and do it in ten to twelve seconds! A suitable lake would be at least two statute miles long and a minimum of six feet deep. This would allow for an engine failure just off the water, after rotating at eighty knots but still below the V1 decision speed of ninety-five knots for water operations. The crew could then land at once following an engine failure and remain within the confines of the lake.

Foam systems were later installed, and pilots could inject a predetermined amount of foam concentrate into the water load immediately after becoming airborne following a scoop. The addition of an onboard foam system increased the fire suppressant characteristics of the drop over straight water. Rather than a rain of droplets from a straight water drop, the foam load had a blanketing effect that smothered the fire and was more effective.

A bonus of the CL-215 in rough water operations was its very robust hull. The CL-215 had one attitude for water-skimming operations. However, if a pilot let the nose get a little higher than normal in rough water on the scoop, the plane would often become airborne prematurely. The pilot could then ease the aircraft back onto the water to continue the scoop. A pilot had to assess whether it was possible to scoop in very rough water. The key factor in rough water for a stable scoop was having the wave crests no farther apart than the sixty-five-foot length of the aircraft's hull. The amplitude of the waves was of secondary concern, with the main feature being that a pilot could feel the variation in the drag from the probe filling the tanks when it went through the wave crests, followed by the gap of air between the wave crests, then over again repeatedly.

Scooping into strong winds resulted in a lower ground (water) speed. This required a longer time on the water to cover the distance to scoop a load. Too rapid a power application and the aircraft would simply fly out of the water before scooping a full load. The opposite was the case with glassy-water scooping. For glassy-water conditions, once the pilot was established on the step with the probes down, the power was promptly advanced to a maximum 53.5″ MP (manifold pressure) to maintain speed during the scoop. It was essential that the pilots confirm that the elevator trim was indicating in the neutral position before touching down, or things could get interesting in short order. As with the Canso, the CL-215 could porpoise if the pilots got the parameters right for it. To avoid that, a rule of thumb was not to touch down above eighty knots to start the scoop and not to bring the aircraft out of the water until a minimum of eighty knots had also been achieved following the scoop.

One sure way to induce a porpoise was to try to salvage a scoop when the non-flying pilot had not advanced the prop levers to "full fine" before touchdown. This error became instantly obvious to the pilots as they advanced the power, and the props were still in coarse pitch. Somehow, the dynamics of then pushing the prop levers forward to their correct position with a resulting surge in power, along with other forces, produced an immediate porpoise. This happened twice in Tim Garrish's twenty-five years flying the machine, and both occasions are etched in his memory. No damage was done to the aircraft following the gyrations in and out of the water.

Considering the many hours of operations over the last fifty-plus years of operations, the water bomber has an excellent safety record in Canada. Operators in Canada adhere to one basic rule that contributes to the safety record: *never bomb uphill.* This ensures a pilot has an escape route to lower terrain in the event of problems. Operators elsewhere in the world do not strictly adhere to this rule, and that may have contributed to several accidents, particularly in Europe, over the years.

• • •

Under the Cooperative Supply Agreement developed by the Canadian Council of Ministers of the Environment in July 1983, the federal government and six provincial governments acquired twenty-nine CL-215s. The federal government purchased seventeen of these (the balance was purchased by individual provinces), which were leased to Newfoundland, Quebec, Ontario, Manitoba, Saskatchewan, and Alberta for a period of fifteen years, after which the title was transferred to the lessees. The lessee provinces were responsible for operating and maintaining the aircraft and making them available to other members. The federal government also bought two each for the Northwest Territories and the Yukon.

The Yukon government relinquished their two CL-215s to the Northwest Territories in 1991, as the latter got more use out of them with the numerous lakes of the Canadian Shield in the Northwest Territories. They could also be used along the one-thousand-mile length of the Mackenzie River, a natural scooping source running by many communities from Great Slave Lake to Inuvik. Each spring, the Coast Guard put channel markers along the river to help the seasonal barging, and, by chance, the marked channel had at least six feet of water, which ensured sufficient depth in the murky, silt-laden waters of the Mackenzie River. The aircraft also carried an anchor as part of their original equipment so, in the event of a problem, the aircraft could be secured from the whims of the current. Other rivers, such as the Liard River and the Slave River, both muddy water rivers, were used, but crews had to be familiar with these specialized operations.

On August 19, 1986, as part of the federal government's plan to sell off state-run corporations to the private sector, Canadair was sold to Bombardier for a bargain $120 million. As an incentive for the mass-transit company to buy it, financial aid was provided to help increase sales of the CL-215. The provinces were offered a free CL-215 with every purchase, while the Northwest Territories and Yukon received two free CL-215s, even without a purchase.

Production was ramped up, and deliveries were increasing both to established and new CL-215 operators. In addition to the Yukon and Northwest Territories, Alberta, Saskatchewan, Manitoba, Ontario, Quebec, and Newfoundland added CL-215s to their fleets in varying numbers. British Columbia chose not to get the CL-215, a decision that was regretted in later years. After 1986, the CL-215 aircraft had become the backbone of Canada's airtanker fleet.

Production of the first eighty aircraft was completed in batches of thirty, twenty, fifteen, and fifteen, and the last CL-215 was delivered to Greece on May 5, 1990.

A turboprop version, the CL-215T, was announced in 1987. It had Pratt & Whitney PW123AF turboprops with engine mounts from a De Havilland Dash 8-300 to provide a 15 percent increase in power over the older and heavier P&W R-2800 radial engines.

The Bombardier 215T had a maximum cruise speed of about 350 kilometres per hour (189 knots) and, in an average mission of six miles from water to fire, could complete ten drops in an hour, delivering twelve thousand gallons of fire suppressant. The first of these prototypes flew on June 8, 1989, and retrofit kits for piston CL-215 aircraft to the new standard were offered.

With the CL-415 in the wings, the company elected not to continue with the CL-215T kits. This gave Cascade Aerospace of Abbotsford, B.C., the

The CL-415s in Italian fire service. When Bombardier closed the production line in October 2015, ninety-five CL-415s were in use around the world.

opportunity to enter the CL-215T conversion business in 2010, converting four of Alberta's CL-215s and four owned by Saskatchewan. Since the federal government had provided considerable financial support to Canadair, it also encouraged (i.e., expected) more resource-sharing of the aircraft to meet fire needs as they occurred across Canada. This was done through the federally and provincially funded Canadian Interagency Forest Fire Centre (CIFFC) in Winnipeg. The role of CIFFC is to ease resource sharing, including aircraft, across the country when fire conditions warrant it.

CL-415

The CL-415 made its maiden flight on December 6, 1993, with the first deliveries in November 1994. The more powerful Pratt & Whitney C123AF engines and the larger composite propellers meant several aerodynamic changes to the aircraft's design. To maintain engine-out controllability, they are mounted closer to the fuselage. The large winglets/canted wing endplates improve directional stability, and an inverted fixed leading-edge slat is mounted forward of the right-hand horizontal stabiliser. The large, highly cambered wing gives the CL-415 satisfactory STOL performance, with a slow stall speed of seventy-eight miles per hour.

Two water tanks, each holding 674 gallons (3,066 litres) of water, located at the aircraft's centre of gravity, are below the floor in the hull, with each side having a header tank that rises into the fuselage compartment. Forward of each water header tank is a 340.5 litre foam tank with enough foam for twenty drops. The foam tanks are usually filled with a short-term retardant that "gels" the water, allowing it to hang up in the forest canopy and delaying run-off when it reaches the ground. Maximum takeoff weight from the ground is 43,800 pounds (19,867 kilograms) while maximum weight after scooping is 47,099.558 pounds (21,364 kilograms).

Water is "scooped" through two hydraulically actuated three-by-five-inch probes, immediately aft of the hull's step.[3] Forward of the step are the four hydraulically operated drop doors, two for each tank.

Peter Marsh related he was happier with the CL-415: "Simplified engine handling," he conceded "modern flight instrumentation, powered flight controls with a rudder trim compensator — all combined to enhance safety and reduce workload. Crosswinds were a breeze. Everyone appreciated the ground level centre point refuelling, and when required, engine oil is replenished from an inside panel. Everyone's favourite improvement was the air-conditioned flight deck. Loads exit their tanks slower, more agitated, and land softer, with a larger footprint. I prefer it, and fire crews appreciate not cutting their way through fallen trees on the fire line. It is hard to find fault with the 415."

In 1998, Ontario purchased nine CL-415s for $225 million (including training and maintenance). This expense was offset through an agreement whereby Bombardier would buy back the existing CL-215 fleet and conduct the final CL-415 assembly in Ontario, creating fifty local jobs. The final CL-415s were thus assembled at Bombardier Aerospace's facility near Jack Garland Airport in North Bay, Ontario. When Bombardier closed the production line in October 2015, ninety-five CL-415s had been completed. De Havilland Canada acquired the Canadair CL program in 2016. Viking

This historic image pays tribute to the very best of Canadian-built water bombers — the Canso, the CL-215, and the CL-415.

Air, through that company, renamed the CL-515 the DHC-515, planning for production and final assembly in Calgary, Alberta. With its certification expected in 2025, the DHC-515 will have modernized avionics, nighttime operational capabilities, and be capable of refilling its tanks from fresh or saltwater in twelve seconds.

Orders have ramped up from Greece, Spain, Portugal, France, Croatia, and Indonesia. De Havilland Canada has announced that the earliest any DHC-515s would be ready for Canadian customers would be 2029 or 2030. To think it all began with OPAS pilot Carl Crossley installing a forty-five-gallon drum in the front cockpit of a biplane!

If you were to choose an aircraft that embodies the very best qualities of your country, for the British it would be the immortal Spitfire; the United States, the trusty DC-3 or C-130; the French, *le plus sexy* Concorde. With Canada, it is the slab-sided, garishly yellow, ugly duckling Canadair water bomber that looks like a toy that escaped from a cereal box.

As to its beauty, another aviation author, Antoine de Saint-Exupéry, said it in *The Little Prince* better than I could: "And now here is my secret, a very simple secret: It is only with the heart that one can see rightly; what is essential is invisible to the eye."

CHAPTER 11

Conair

CONAIR IS THE LARGEST FIXED-WING, PRIVATELY OWNED AERIAL FIREFIGHTing company in the world. Proudly Canadian, it operates out of Abbotsford, British Columbia. Its fleet includes the latest birddogs, land-based airtankers, and amphibious aircraft. Its Training and Tactics Centre, with simulators that create missions in real-world operations, is the envy of other companies.

It all began fifty-six years ago when Leslie George "Les" Kerr joined Art Seller's Skyway Air Services in 1952 as an agriculture spray pilot. In 1969, Kerr and Barry Marsden bought the company's Air Services Fire Control Division, purchasing from Seller thirteen Grumman Avengers, five Boeing Stearmans, a North American Harvard, and Skyway's hangar at Abbotsford.

Marsden remembered how the name "Conair Aviation" came to be. "We were brainstorming; the lawyer was waiting on a name to finalize the paperwork, and we'd been around and around, until he prompted us with a reminder of our work in crop spraying and firefighting. We said, we're in the business of aerial contracting and aerial control, and so the name Conair was born."[1]

When he retired in 1995 as president and CEO of Conair, Kerr had overseen the operation of the company's fifty fixed-wing aircraft, forty helicopters, and four divisions: Conair Aviation Division, Conair Europe, Frontier Helicopter Division, and Aerospace Division.[2]

Conair expanded its interests in the United States in 2010 by purchasing Aero-Flite, Inc., an aerial firefighting operation based in Kingman, Arizona. They relocated to Spokane, Washington, in 2015, with their CL-415s and the land-based RJ85 and Dash 8-400AT airtankers.

In the late eighties, Ray Horton transitioned to airtankers, initially flying the Firecat and then the DC-6. Both airtankers had mutated from earlier aircraft in other roles, the former from the Grumman Tracker and the latter from the last of the great piston engine airliners.

Grumman Tracker

The Grumman Tracker was a stubby, highly manoeuvrable aircraft originally designed to operate from aircraft carriers to track foreign ships and submarines. It was a twin-engine aircraft, powered by Wright Cyclone R-1820 piston engines, each delivering up to 1,525 horsepower. With a rapid roll rate and for the most part good low-speed handling characteristics, it was a popular aircraft for firefighter conversions. The aircraft had a fixed slot in the leading edge of the wing ahead of the aileron and a mechanically operated spoiler, which deflected in unison with aileron input for banking/turning the aircraft. This allowed for particularly crisp and quick bank angles when commanded by the pilot; however, as it turned out, these devices also created a very nasty characteristic. At slow speed with high power on the engines, the Tracker tended to suddenly stall, typically pitching the nose very high, through the stall angle of attack and into a rapid roll, quickly inverting the aircraft.

The slot/spoiler combination in the outer part of the wing provided excellent handling characteristics right down to the stall speed, allowing for the same crisp handling at low speed as it did at high speed, with no perceptible airframe buffeting or sluggish handling common to most other aircraft near the stall angle-of-attack. However, the inner part of the wing near the root did not have such devices and could reach the stall angle of attack well before the outer portion, which, while still providing reasonable airflow over the ailerons, masked the impending stall to the pilot. While such stalls were recoverable at higher altitudes, this was deadly at the low altitudes flown by airtankers while firefighting.

Declared surplus to the military's needs, Trackers were converted into Firecat airtankers by Conair. When the French government's Sécurité Civile began its "Guet Aérien Armé," or "armed aerial war," in the mountainous south of their country in 1980, it looked for a fast airtanker with a three-thousand-litre capacity that was more cost-effective than a Canadair water bomber. The following year, France ordered three Firecats from Conair, the first of fourteen over the years.

Firecat

The Conair Firecat carried 3,296 litres of retardant in an internal tank, where the original navy torpedo bay had been located. The bottom of the tank was aerodynamically flush with the belly of the airframe and consisted of four independently operated doors running lengthwise. The doors could be operated individually for four separate drops, two at a time, four at a time, or individually one after the other to create a "string," or lengthened drop. The type of drop was typically determined by the size of the fire, vegetation type, etc.

SMI

Conair lost three Firecats with their pilots within the first ten years of Tracker operations, the first one occurring with the very first airframe during a demonstration flight. During the same period in the 1980s, the French lost eight Firecats to stall/spin accidents while firefighting.

Following the third Conair crash in 1988, the company commissioned an engineering program together with an external provider to design and develop an angle of attack instrument known as a stall margin indicator (SMI). This was installed in the pilot's field of view on the glareshield and provided advanced indication of a high angle of attack that also gave protection from high "g" loading stalls often encountered during steep angles of bank or accelerated nose pitch-ups commanded by the pilot. The SMI transformed the way the aircraft was flown, and the company never encountered another Firecat stall/spin episode or associated loss of airframe and life through to the aircraft's retirement in 2012. This simple instrument was so successful that Transport Canada and the Federal Aviation Administration (FAA) mandated that they be installed in all S2 Trackers used for firefighting. Horton flew the Firecat both before and after the installation of the SMI and experienced firsthand how transformational this instrument was to the way he had been flying the aircraft from takeoff, through firefighting, and back to landing. He knew then how close he had been to becoming a statistic before the SMI installation.

When the French wanted an upgraded version of the Firecat, Conair developed the Turbo Firecat. It featured Pratt & Whitney Canada-powered PT6A-67AF turboprop engines housed on cowlings constructed of lightweight composite materials, replacing the heavier piston engines. It also had five blade propellers, a larger tank (3,395 litres), a foam injection

Conair built Firecats for the French government firefighting agency, Sécurité Civile, and later converted them with P&W Canada–powered PT6A-67AF turboprop engines.

system, and a single point refuelling system. Underwing fuel tanks allowed the aircraft to be reloaded and refuelled in less than four minutes. After forty years in France, Conair's Firecats marked 2020 as the end of their service fighting wildfires.

DC-6 Airtanker

Conair began operating the DC-6 in 1970, eventually amassing a fleet of ten airframes. The DC-6 was designed and built by the Douglas Aircraft Company in the early 1950s as a long-range airliner. They were flown by many airlines to inaugurate long-haul oceanic routes as far away as Australia.[3] While originally designed with 107,000 pounds gross weight (48,534 kilograms), Conair's DC-6s were reduced to 103,800 pounds gross weight (47,082 kilograms) for cargo hauling and 97,000 pounds gross weight (43,998 kilograms) when firefighting to allow for safer low speed handling and "g" loading.

Being a few years ahead of the jet-age, the DC-6 enjoyed a relatively short airline passenger life, with most retired from airline service by the late 1970s and early 1980s. With Conair's DC-6 operations beginning in 1970, these aircraft were relatively young compared with other converted firefighting types of world-war vintage, with some only twelve years old!

Their engines, developed specifically for many hours of long-range flying and few power changes, were very reliable in the airline world but tended to be temperamental in more torturous firefighting operations with numerous power changes. Ray Horton lost count how many times he returned to base with an engine feathered due to cylinder problems.

The "6," he recalled wistfully, was "a dream to fly. A very stable platform, it was reasonably manoeuvrable for its size, and it easily transitioned to its new role as an airtanker. The aircraft had a thick leading-edge wing, giving very traditional flying qualities that provided ample warning with significant airframe buffeting ahead of an aerodynamic stall. At the point of stall, it had no unusual characteristics and was easily recoverable with minimum altitude loss. The ailerons remained effective throughout the stall.

The airtanker modification consisted of fitting an external retardant tank that hung below the belly of the airframe capable of carrying 11,356 litres

One Remarkable *Six*

Conair's C-GIBS

The Conair DC-6, a very stable platform reasonably manoeuvrable for its size, easily transitioned to its new role as an airtanker.

(2,500 gallons). The tank was initially equipped with eight external doors, with two sets of four doors arranged longitudinally from left to right across the bottom. Later, Conair developed an updated twelve-door version, with two sets of six doors, each allowing for more versatile use of the same retardant load. The doors operated the same as the Firecat, with multiple drop options, and had an electronic drop selector.

The DC-6 is a large aircraft and as such required good planning when flying low level in mountainous terrain and executing firebombing runs. It was by no means over-powered, and thus DC-6 pilots had to exercise extreme caution before descending into a mountain valley for a drop to ensure the aircraft had a good and safe escape route back to altitude again in the event of an engine failure or retardant tank failure. Pilots could not rely on being able to outclimb the nearby hills, especially in the hot summertime temperatures. This was where the birddog pilot would be heavily relied upon to prefly the circuit and identify any significant hazards within the proposed drop area and provide explicit instructions on the escape route to fly. In mountainous terrain, the birddog pilot often had to fly many miles along a valley to ensure it would lead to a safe escape and not outclimb a crippled DC-6 airtanker. If so, another exit route would have to be found.

Flying the DC-6 close to the ground with its minimal performance could be an unforgiving endeavour if pilots didn't provide their full, undivided attention to the terrain around them or didn't follow the preflown route provided by the birddog. To that point, on August 2, 1974, Conair's chief pilot at the time held his load and initiated a go-around when he was offline at the drop point. He elected not to follow the exit routing he'd been given and turned the opposite way to shorten his return for another drop. Unfortunately, the terrain in that direction was slightly rising, something the birddog pilot had noticed from his numerous dummy runs while waiting for the airtanker to arrive, but the DC-6 pilot had not. Despite many warnings from the birddog pilot to drop his load (to gain performance from the lightened load), the DC-6 captain attempted to complete the turn while keeping the load, until the aircraft contacted the ground with all crew lost. Many lessons were learned from those early days, and Conair never lost another DC-6 following that accident through to its retirement from the fleet thirty-seven years later.

In the older airtankers, to reduce weight and provide as much capacity as possible for retardant, luxuries such as autopilots, pressurization, lavatories, passenger seats, and even interior walls and insulation were removed from the aircraft during conversion. The DC-6 was no exception, but it did not detract from its flying qualities. As some of these items are appropriately considered fatigue-reducing aids today, most modern airtankers now maintain functioning autopilots and air conditioning/pressurization systems.

"Nevertheless, in the 1980s and '90s," Horton said, "the DC-6 was the queen of our fleet and formed a large part of the B.C., Alberta, and Northwest Territories airtanker fleets. Those of us lucky enough to fly it were a privileged lot for sure."

Even before the availability of 100LL fuel became a crucial factor, and far ahead of other airtanker companies, Conair made the transition from piston-engine aircraft like its DC-6 and Tracker to turbo prop L-188s and Convair 580s. The P&W R-2800 radial engines on its Convair 580s were replaced with more powerful Allison 501 turboprops, the 1940s airframe lightened by removing pressurisation and flaps reduced for a belly tank. In a long-term strategic move, Conair then replaced the L-188s and Convair 580s with De Havilland Canada DHC-8 Q400s. The company's pilots were said to be glad to get rid of the Convair with its cumbersome controls. The saying among them was "Forget the sim, go to the gym."[4]

Dash 8-400

Conair's next choice for a fire bomber was the Dash 8-400, an airliner that unlike older aircraft was fully supported by its manufacturer, De Havilland Canada. The Dash 8-400 had the lowest fuel burn per gallon/litre hauled, making it less costly to maintain, and it could land on shorter runways than other large airtankers, closer to where fires were. External water/retardant tanks were installed on the Dash 8-400 instead of internal ones to preserve the integrity of the airframe, allowing the aircraft to remain pressurized and reducing pilot fatigue. The external configuration also enabled the multirole variant of the Dash 8-400 airtanker to have its tank removed during non-wildfire emergency response missions transporting cargo and passengers. As Conair president and CEO Matt Bradley explained, "For example, we can change from an airtanker to a passenger/freighter in two hours or to a full medivac configuration in three hours." In 2023, Conair was turning out the Dash 8-400MRE and one Dash 8-400AT airtanker every seventy-five days.

BAe RJ-85

In 2013, Ray Horton helped launch the RJ-85 airtanker aircraft into Conair's American subsidiary, Aero-Flite, in Spokane. He said,

> I was fortunate to be involved with Conair's search for a "Next Generation" airtanker in the mid-2000s. We evaluated several aircraft, including the C-130 Hercules and

> Boeing 737-300/400/500 series, as well as the BAe 146-200. Although Conair had already converted the Dash 8Q-400 into an airtanker for the Government of France, the Q400 at that time was enjoying a heyday with the airlines and was still too expensive to acquire and compete competitively for customer contracts. The used aircraft market changed dramatically during COVID, which allowed the Dash 8-400 to find much success with Conair in the competitive market throughout the world.

To choose the next generation of airtanker, Ray Horton explained,

> We developed a matrix and evaluated the same parameters for each aircraft type we identified as possible contenders. The parameters identified were extensive and included things such as manoeuvrability, slow speed handling, projected load capacity, cost to acquire, projected modification and certification costs, and, very importantly, original equipment manufacturer (OEM) support for the modification. Other parameters included ongoing airframe and parts availability, maintenance costs, and future OEM support for the airframe, to name a few.
>
> As can be imagined, converting an airplane such as an airliner into a competent airtanker is a hugely complex and costly affair and is subject to full regulatory (Transport Canada) oversight combined with an extensive flight test program before certification can be granted. The engineering required not only to design a functional and competent retardant delivery system (RDS) that meets the very strict drop-standard requirements, but also to structurally attach it to an airframe originally designed for a different purpose, is a massive undertaking from an engineering perspective. Of course, this must be accomplished with zero or minimal disruption to the original flight and handling

> characteristics of the aircraft. To remain cost-effective, the modification must also have minimal impact on the maintenance programs developed for the aircraft by the manufacturer for its original role.
>
> Complex indeed, and as can be seen, OEM support is crucial for access to original engineering development and assembly data. This is highly guarded information, and many OEMs may not want to share it. While all the aircraft we evaluated had many positive attributes, the BAe 146 ranked highest on the matrix, with BAE being the most supportive OEM of our program and agreeing to share all engineering data and support the program through certification with Transport Canada and the FAA and beyond.
>
> During our extensive evaluation and discussions with British Aerospace, we determined the second generation of the 146-200, known as the RJ85, was a better fit for our program, since many beneficial upgrades had been taken by the manufacturer with this later generation. Some of these included updated engines and flight deck instrumentation and a higher level of automation. Most importantly, while more expensive, the RJ airframes were much younger than the 146s and therefore provided a much longer life expectancy.

Ray thought the aircraft was a dream to fly. He said that one BAE test pilot commented during his first flight, "Are you sure the tank is even attached?" Even with the external RDS tank attached to the belly, there were no discernible flight characteristic differences from the unmodified airliner. In fact, there were no changes or restrictions imposed to any of the original certification parameters of the aircraft. This is highly unusual in such an extensive airframe modification program.

The RJ retardant tank developed by Conair carries 11,356 litres (three thousand USG) with excellent flow rates, which allows for a clean, wide drop and provides highly effective retardant lines on a fire's edge. The tank has an electrically controlled, hydraulically actuated door system that provides a constant flow of retardant for the selected quantity, together with multiple drop choices, from as little as one-tenth of the load per drop right up to a full load per drop. Various coverage levels from 1 to 8 are selectable, which equate to thickness of retardant as it is applied to the different fuel types (1 being thin, used in light fuels such as grasses, 8 being thick, for heavy timber). The pilot interface or tanker controller is simple and easy to use with all system annunciator lights located on the flight deck glare shield in the direct view of both pilots. This allows for minimal eye movement while the crew concentrate outside the aircraft during low-level manoeuvring and firefighting runs.

The RJ85 is a very satisfying aircraft to fly from a pilot's perspective. Airliners are generally optimized for high-altitude flight with automatics engaged. The airliner version of the RJ85 generally flew at slightly lower altitudes and remained very responsive with its automatics off — mandatory for low-altitude firefighting. It's exceptional slow-speed handling allows for easy manoeuvrability around a fire while providing good feedback through the controls to the pilot. During the RJ modification program, Conair developed and certified a Flight Envelope Awareness System (FEAS), which provides even more awareness to the flight crew of real time manoeuvrability loads and safe margins on the airframe. Together with advanced crew training, this contributes to the very safe and efficient delivery of retardant to the fire.

The RJ85 is a true all-weather capable airtanker, with the right level of automation and certification to safely transit to and fight fires anywhere in the world.

Conair, with almost military precision, has progressed from strength to strength through the decades, in Canada, the United States, and in Australia. With the impact of global warming on wildfire in the decades ahead, one can only hope it will continue do so.

CHAPTER 12

Unsung Heroes

BEFORE THE ARRIVAL OF FRESH-OFF-THE-ASSEMBLY-LINE CANADAIR WATER bombers, private contractors in the United States relied on retired military aircraft and outmoded airliners in the war against wildfires. France, Spain, and other European countries had fleets of tankers that were owned and maintained by their governments, but the United States did not. Repurposing fifty-year-old aircraft to perform exercises never envisioned by their manufacturers would reasonably have demanded continuous monitoring of their structures to ensure the safety of the crew. Not so in the United States, where, as "public use aircraft," the tankers were at the time exempted from the Federal Aviation Regulations that governed all other aircraft.[1] As such, the owners of the flying museum pieces were not required to have a continuing airworthiness program. Their safety oversight was the responsibility of the Forest Service, which, as part of the Department of Agriculture, had no expertise to provide it.

Private operators were not required to use black-box flight recorders, and, until 1996, the National Transportation Safety Board (NTSB) had no authority to investigate airtanker crashes. Crucially, there was no support from the military or the aircraft's original manufacturer, sometimes no longer in existence.

These operators were not given to spending a lot on the servicing of their fleets either. Living hand to mouth from summer to summer, the airtanker companies inhabited a world in which the Forest Service used a single criterion when selecting aerial firefighting contractors, i.e., the lowest cost. Fire season contracts were renewed annually at the time, and canny tanker companies ensured that their costs were always less than the Service's contract offers, or they would be out of business. If they undertook the expensive safety modifications for their decades-old surplus aircraft and did not get the bid, the airtanker companies knew they would go bankrupt.

As expected, stressed air resources mixed with an increasingly aggressive wildfire environment resulted in several fatal aerial firefighting accidents. These rarely made the news, until the 2002 fire season. On June 17, 2002, a repurposed Lockheed C-130A Hercules, N130HP (retired to the boneyard by the military in 1978), experienced an in-flight breakup begun by the separation of the right wing, followed by separation of the left wing. Its two crew were killed. On July 18, 2002, a dinosaur of a Consolidated Vultee P4Y Privateer, N7620C (built in 1945 and retired from the military in 1956), lost its left wing while manoeuvring to deliver fire retardant over a forest fire near Estes Park, Colorado, killing five aircrew. In both cases, no maintenance or flight-hour records were available.

Between 1958 and 2002, 136 airtanker pilots had died in the line of duty. But in 2002, both tragedies were captured on video by television crews and played repeatedly to horrified audiences on national television. As happened in the "Big Burn" of 1910, the 2002 air crashes came at a critical time in United States history. It took the telephone seventy-five years to reach fifty million users; it took television thirteen years. But it took the World Wide Web just five. Following the 9/11 attacks, Americans were very sensitive to images of air crashes, and when video of the accidents spread across the internet, the response was immediate and strong across the nation. Suddenly, "fracture fatigue," "tanker widows," and the Airtanker Scandal of 1987 made the news.

Like the Storm King Mountain tragedy in 1994, when fourteen firefighters died, the two crashes triggered long-overdue changes in aerial firefighting strategies. The Forest Service and the Bureau of Land Management jointly established an independent commission to plan a safe future airtanker program.[2] The commission released a scathing report in 2004, which caused the Department of the Interior and the Department of Agriculture to rescind all contracts for heavy airtankers. Only those companies that met the new safety-related obligations were accepted.[3] Bankruptcies for many airtanker companies followed, a few hanging on until their contracts were reinstated in 2006.

The commission concluded that an overhaul of the system was vital before more lives were lost. Both Canada and, in particular, Conair were cited by the commission as models for the United States to emulate. Accelerated-aging factors had long been used in Canada's firefighting aircraft airworthiness certification and maintenance protocols. One firefighting flight hour might be equivalent to seven "normal" flight hours, and this significantly shortened inspection and structural maintenance intervals and the aircraft's life. Canada also certified its firefighting airtankers in the Restricted category but went far beyond any U.S. criteria. Transport Canada had special conditions to cover special-purpose operations as part of its large aircraft certification in the Restricted category.

• • •

Conair was recognized by the commission as an example of how Canadian operators complied with these requirements. The company had developed a Supplemental Structural Inspection Document in 1996 for its Fokker F-27 firefighter aircraft and then submitted it to Transport Canada's Continuing Airworthiness Aircraft Certification Branch chief for approval. According to the Conair plan, its F-27's standard structural inspection time intervals and limitations accounted for the increased severity of the time spent in the firefighting role as compared with typical Fokker F-27 transport role operation. The company had developed a set of formulas to calculate the degree that an aircraft's lifetime was reduced when flying the fire retardant–delivery portion of an airtanker mission.

"To understand the unique criteria required to modify an aircraft into an airtanker, one must consider the different flight environments," explained Michael Benson, director of business development at Conair Aerial Firefighting. "For example, take a civilian aircraft fully loaded with passengers and enough fuel to make a one-hour flight. It takes off, climbs to twenty-five thousand feet, it is pressurized and cruises in roughly a straight line. Just prior to arrival, it depressurizes while it descends for landing. The aircraft is assumed to be subjected to average turbulence and relatively minor flight manoeuvres. This is a very typical flight profile for a certified civil aircraft," he said. "With the science of fracture mechanics, cracks in metals grow at a predictable rate when the magnitude of the load and the number of cycles the aircraft has been subjected to are known. When an aircraft is flown within the defined safe flight parameters for which it was certified, a straightforward maintenance program is implemented. This civil flight environment is predictable, and so is fatigue damage."

Benson continued,

> But if the same aircraft is converted into an airtanker responding to a wildfire, it takes off with a full tank of

The Convair 580's lower wing skins required constant inspection and repair, as the airframe had been initially designed for radial engines.

> retardant and enough fuel for three hours of flight. The airtanker climbs to twenty-five thousand feet, it descends about half an hour later to three thousand feet — partially depressurizing, it flies over a forest fire in a relatively gusty environment. It then performs a series of low-level manoeuvres and descends to 125 feet above ground level. It performs a retardant drop, instantly losing 30 percent of its total weight; this is followed by a pull-up and bank manoeuvre, then the airplane ascends and repressurizes to twenty-five thousand feet. It then returns to land for a reloading of retardant and possibly to refuel. The firefighting cycle involves multiple pressurization cycles, more low-altitude manoeuvres, and is exposed to a gustier environment. The typical aerial firefighting cycle flown by a tanker is more damaging to the airframe than the typical civil cycle flown by a commercial aircraft.

Rates of descent during retardant-drop runs range from zero for flat approaches to nine thousand feet per minute during steep approaches in canyons or along canyon walls. The latter figure is comparable to what an F-16 fighter aircraft might achieve during a steep-angle bomb delivery. Pullout load factors on the tanker during retardant drops and recoveries range from 0.5 G to 3.9 G. Hence, diligent maintenance is essential.

"If the airplanes and pilots are the heartbeat of the operation, then support staff are the bloodline," said Conair's RJ85 airtanker captain Anthony Ussher. "Typically, the relationship between aircraft maintenance engineers (AMEs) and pilots is detached. But here at Conair, the relationship is far more dynamic. We make most decisions together. We are both invested in the outcome, and honestly — they know the plane best."

As Conair AME Megan Nunner explained,

> Hearing a pilot report that the aircraft is running perfectly is music to my ears, a testament to the team's hard work and dedication. I've worked on many different types of aircraft,

> but primarily Convair CV580 and Dash 8-400AT. I'm away from home for four months out of the year on base during the fire season. Life on base is very different from my day-to-day work life. Each day on base, my main goal is to ensure we are ready to always support the aircraft. It's an emergency environment, and the aircraft must be ready to dispatch when needed. Sometimes this means you end up working overnight on the airtanker, as it operates during the day on wildfires. No day is ever the same.
>
> The reason I love my job is knowing that it contributes directly to safeguarding communities and forests — and that provides me with a deep sense of satisfaction and pride. I would recommend this industry to any woman wanting to join. The job satisfaction at the end of the day is priceless.

A former director of maintenance, Ray Bozzer provided support for more than forty-five years in the aviation industry. He unravelled the intricacies of maintenance to me in his own words:

> Aerial firefighting aircraft have very specialized requirements and diverse needs for different situations. Each tanker type has its Achilles heel, and there are specific considerations required to keep them in the air. For airtankers and birddogs, there are four phases of maintenance:
>
> **Line Maintenance**
>
> Line maintenance is conducted at the tanker bases and usually performed before or after alerts. These are periodic checks, calendar checks, and "out of phase" tasks. On older tankers, such as the Douglas A-26 or DC-6, the inspections were all periodic, typically every fifty hours. In the past, this was simple to follow, as an AME could keep track of everything on a small chalkboard. Some operators

developed a seven-day/twenty-five-hour inspection for the older aircraft to ensure that the aircraft was being inspected during times of inactivity. During the seven-day inspection, the AME is looking for fuel, engine oil, and hydraulic leaks, accumulator, and tire pressures, the function the retardant delivery system and flight controls, damage to the aircraft from vehicles, stands, as well as birds' nests. When there is a fire "bust," the aircraft needs a once-over every twenty-five hours.

Typically, Conair AMEs also walked around the aircraft when they were being reloaded in the retardant loading pits, looking for loose exhausts, any damage to the aircraft from bird strikes, and leaks. Extreme caution is needed when carrying this out, as the engines are running, and the ramp could be slippery from the retardant.

With the newer airtankers, computer tracked forecasts are sent out to the AMEs on a regular basis to ensure that none of the tasks are "over-flown."

Preservation

When the aircraft returns from base, the AMEs perform engine evaluation runs, preserve the engines, remove the survival gear and other equipment. The retardant delivery system is washed, and in some cases, depending on what maintenance is done, it is removed.

While washing the aircraft, the AMEs are looking for any anomalies, such as missing fasteners and cracks, and noting them down so that they can be repaired. With the older aircraft, such as the A-26 Invaders, a returning group would consist of three tankers and a birddog. The A-26 was designed by Douglas during the Second World War as a medium bomber with a short-projected life in combat, and ongoing maintenance was not factored into the design.

That made some of the tasks awkward to carry out. Going out into the field with a group of three Invaders, you were guaranteed to change one engine, half a dozen cylinders, four magnetos, a dozen exhaust pipes, a couple of brakes, and main wheel assemblies. That was above the routine maintenance. The AMEs really earned their keep by keeping the aircraft ready during alerts.

Winter/Heavy Maintenance

Months in advance, a forecast is created, and a tentative maintenance plan is created for each aircraft. Every aircraft has its own maintenance plan, but many of the tasks are the same, such as function checks of the landing gear, flight controls, hydraulic system, etc. Once the aircraft has been preserved, a detailed custom maintenance plan will be created for each aircraft. This is when all the major systems and structures are inspected in greater detail. Structurally, the AMEs are looking for cracks, corrosion, lose or missing fasteners.

Built during the war, there wasn't much corrosion prevention done on the A-26 Invader, so a lot of work went into treating the corrosion and its prevention. With the Firecat, it was the welded skins. Grumman decided to weld the skins to the strings. Normally, they are riveted together, but Grumman felt it would be better to weld. Since the aluminum is bare for this process, corrosion would start in areas that require a lot of effort to inspect and repair.

The Lockheed Electra L188 has its wing plank challenges. Due to the material that Lockheed used with the machining processes of the day, the wing planks are prone to cracking. The other issue with the Electra is that Lockheed is no longer supporting the aircraft, and neither is Rolls-Royce supporting its 501-D13 engine.

The Convair 580's lower wing skins and rear spar caps require constant inspection and repair because the airframe was initially designed for radial engines.

Though CL-215 scoopers were the only aircraft specifically designed for aerial firefighting, they had their issues. With the number of water drops and landings during a fire season, the front and rear lower spar caps and the lower wing skins had to have a detailed visual inspection annually, and every other year, depending on how much flying is done, a nondestructive inspection (NDI) must be conducted. The CL-215T, CL-415, and the CL415EAF also have the same issues.

With the newer airtankers (Q400, RJ85, B737), there is a significant amount of nondestructive inspection, in addition to the other inspection tasks. High-frequency eddy current and ultrasonic inspections are the primary methods, followed by radiography (X-rays).

When the maintenance and modifications have been completed, the aircraft will have the engines preserved again and its airframe and engine plugs installed. It is towed to the ramp, where it sits awaiting activation and possibly flight-crew training.

There will also be a test flight program where the aircraft is instrumented to see if the retardant tank has any adverse effects to the flying characteristics. Another part of the flight test program will be how releasing the retardant, commonly referred to as "the load," will affect the aircraft. Ideally, the aircraft should continue to fly straight and not pitch upward. Once everything has been proven, the regulatory body will issue the Supplemental Type Certificate (STC).

Activation

Here, the aircraft has all the preservation oils drained out, components are reinstalled, it's fuelled up, and the engines are run. All the regulatory items must be complied with, such as pitot-static leakage tests, transponder, and emergency locator transmitter (ELT) checks. After activation and the documents are signed by an endorsed AME, the logbook is signed out. The flight crew takes the aircraft out for a maintenance flight to ensure that all systems function properly. The flight is between an hour or two, and many of the "bugs" will have been worked out by then.

Improvements

The Flight Envelope Awareness System (FEAS) is a modification that enhances the loads monitoring and angle of attack (AOA) systems. This system gives the flight crew advanced warning of a potential stall. The FEAS gives both aural and visual warnings of an impending stall.

Other

There is a lot of camaraderie at the tanker bases. In the past, forest service personnel, support staff, AMEs, pilots, and others would have two meals together every day. The team can be a very close-knit squad. Everyone looks out for everyone else. Normally, it is the pilot's responsibility to fuel their aircraft, but during a "fire bust," he must get a drink and a bite to eat. So, the AMEs will step in to help, even though they are busy doing the other servicing and giving the aircraft the once over. On the flip side, when the AMEs are repairing the aircraft in the scorching sun, or late into the evening, the pilots will bring out food and drinks as a gesture of thanks. Everyone pitches in to keep the machine rolling, with the common goal of keeping the fire at bay so the ground crews can put the fire out.

Matt Bradley, Conair Aviation's current president and CEO: "There is an unquestionable bond that exists between emergency responders, a camaraderie that grows when you live and work together on remote bases, far away from friends and family for months during a year.... We work hard. And we play hard."

Matt Bradley, Conair Aviation's president and CEO, agreed:

> At its roots Conair was a family business. Barry, with his wife, Karen, nurtured the culture into what is today. There is an unquestionable bond that exists between emergency responders, a camaraderie that grows when you live and work together on remote bases, far away from friends and family for months during a year. We trust each other, motivate each other, and support each other. We work hard. And we play hard. Christmas parties, BBQs, base dinners — these traditions are still honoured today. As much as we are disciplined to

execute safe and effective aerial firefighting missions, we value the importance of gathering as a Conair family. You skip the get-togethers, you skip getting to know the details of people's lives. And at the end of the day, that is what counts, all being able to get together again after a mission.

CHAPTER 13

Helicopters and Air Tractors

AFTER THE HEAVY TANKER CANCELLATION IN 2004, THERE WAS A CLAMOUR for helicopters and the quick, nimble single-engine airtankers (SEATs), both of which could zero in for a fast drop of eight hundred gallons (three thousand litres) with pinpoint accuracy.

With the Wildland Urban Interface — where suburbs encroach into undeveloped wildland — growing in the United States alone at two million acres (809,371 hectares) annually, fast precision water drops are vital. Helicopters and SEATs have flexibility of movement, speed of loading and unloading, and quick turnaround times. Unlike the legacy airtankers, SEATs are economical to build, buy, and maintain.

The success of an initial attack comes down to how fast you get your resources to the fire, one SEAT pilot explained to me. For a water bomber that reloads in a lake or the sea, it means the distance they need to travel between the water source and the fire. That point can make or break the success of the mission. A SEAT can reload in ten to thirty seconds and if close to the fire can drop loads every one and a half to two minutes. It doesn't take long for the gallons to add up.

Helicopters

Autogiros had been used in fire detection in the 1920s, but like fixed-wing aircraft, they couldn't hover. In 1943, when Igor Sikorsky built the world's first production helicopter, a writer in the *British Columbia Lumberman* anticipated future developments in fire suppression, suggesting that the invention would provide a safer platform for dropping firefighters to the ground than the airplane. If confronted with dense growth, they could simply transfer to a branch of a convenient tree and climb down.

On June 15, 1943, Frank MacDougall, deputy minister of the province of Ontario's Department of Lands and Forests (DLF), asked George Ponsford, the OPAS chief, to find out if the Ontario government would purchase a helicopter for postwar forestry work. The federal Department of Munitions and Supply was quick to reply to Ponsford on July 1, 1943: "There is no possibility whatever of you being able to procure a Sikorsky helicopter at an early date. All production on that ship is booked for military needs for a long time ahead."[1] When Ponsford then dutifully contacted the Sikorsky Company, he was told to wait until the war had ended.

The first use of helicopters in firefighting began on August 5, 1947, with the Bryant Fire in Angeles National Forest, California. Two Bell Model 47Bs owned by the Armstrong-Flint Helicopter Company arrived and with a portable pump carried water from nearby Big Tujunga Creek to wet down the landing zone, and with the hood of a Jeep as the base of operations, the first heliport was ready. There was no thought of carrying passengers or water — the early helicopters were for reconnaissance, scouting, and mapping only.

The distinctive goldfish-bowl Bell 47 was the first civil helicopter to enter production. Sadly, the expectations of its godfather, Lawrence Bell, of a postwar civil aviation boom with heliports on each street did not occur. It was the U.S. military's use of the Bell 47 in the Korean War for their mobile army surgical hospital units that gained it newsreel (and later television) attention.[2] A helicopter's limited range and passenger-carrying capacity, as well as the expense, made the Forest Service view available Bell 47s and Sikorsky R5s as unsuited to firefighting operations.

The helicopter made its firefighting debut in Canada quite by accident. On June 26, 1946, Inco, the International Nickel Mine Company, hired a Bell 47 to test out an experimental magnetometer on mineral surveys in Sudbury, Ontario. With a neighbourhood wildfire raging, Jack Dillion, the local forest protection officer, had Bell's experimental test pilot, Gerald Demming, take him free of charge to reconnoitre the fire's perimeter and even land near a problem area.

The British Columbia Forest Service was also considering the practicality of using helicopters for moving firefighters into the mountains. This encouraged three former RCAF officers, Carl Agar, Barney Bent, and Alf Stringer, to buy a Bell 47-B3 in 1947 and begin Okanagan Helicopter Services. The first commercial helicopter operation in Canada, the company initially survived on mosquito-spraying contracts.

The Wheeler Springs, California, wildfire in 1948 showed what helicopters could do. The fire began in the Los Padres National Forest and swept across the Sierra Madre range. For the first time, Bell 47s delivered men to the fire line. Thirty-five were airlifted in as many minutes, when walking there, it was said, would have taken hours.

The cautious OPAS purchased a Bell 47B in 1953, crewed by Spartan Air Services pilots, and leased a Hiller 360 from Kenting Aviation. Both had insufficient power to carry much, and when their water-filled wax paper bags were dropped on a controlled forest fire, they burst and spread embers along the fire's edge. The Bell 47B was then fitted with water tanks on either side of its fuselage and a nozzle attachment to spray the water, but it proved ineffective.

"If you are in trouble anywhere in the world, an airplane can fly over and drop flowers," Igor Sikorsky said, "but a helicopter can land and save your life."[3] Where the life-saving potential of the helicopter would feature was in the Higgins Fire on August 4, 1961. The Forest Service had sent twenty smokejumpers to what looked like from the air a routine fire in the Nez Perce National Forest, Idaho. But an afternoon cold front turned the blaze into an inferno, and the smokejumpers were soon surrounded by fire, trapped in the middle of it with no escape route. They dug in, buried their faces in wet bandanas in the dirt, and tried to find air to breathe. The fire was so hot that canteens of water near the jumpers started exploding. The cavalry arrived in the form of Johnson Flying Service's helicopter pilot

Rod Snider, who managed to land his Bell 47G-3B on Higgins Ridge. Rod told them he would take four out on each trip. Two rode in the cabin and two hung on to the cargo trays. Despite heavy smoke and wind, in five trips Snider was able to ferry all twenty jumpers to safety.

In an oral history interview many years later, he recalled, "It was hard to find them. The wind was really cooking in there, and you couldn't see the heliport all the time to get down. I had to come in high and drop down into it when I could see a little break."[4]

"What made you risk your life to do it?" the interviewer asked.

"Oh, it had to be done. It had to be done," Snider replied with characteristic humility. "I don't know. You just can't leave guys down in the position that they were in."[5]

Helicopters transported firefighters to the fire, and experienced staff could patrol fire lines in them, looking for "hot spots." But like the H-Boats of the 1930s, they lacked the power to carry water to a fire.

In 1958, the U.S. Marine Corp tested Sikorsky S-58 helicopters with 150-gallon experimental tanks under their fuselages. They came in time to be used in the Angeles National Forest fire. The same summer, Okanagan Helicopters retrofitted its S-58 with an internal 270-gallon tank under the fuselage.

The search for a practical water bucket that could be slung under and tipped over the fire progressed through the years. Dominion Helicopters of Ontario was the first to experiment with its H-21, lifting a square metal bucket that carried 258 gallons (976 litres) of water under the fuselage. A rope tipped the bucket for filling and dumping the water. Using a lighter Hiller 12-E, Okanagan Helicopter pilots Jim Grady and Henry Stevenson developed a fifty-four-gallon (204 litres) "Monzoon" bucket that could be filled while hovering over a convenient water source. Later, the bottom hatch was activated by a solenoid to release the water.

The "Grady" bucket that followed was a forty-five-gallon barrel that could be dipped in a water source. Once on target, the pilot would release the top of the barrel, allowing it to tip over onto the fire. The "Water Buoy" bucket was an improvement; with it, water could be released pneumatically through doors in the bottom of the bucket. The larger "Griffith" bucket could also be pneumatically operated but couldn't be collapsed for ease of transport.

Then, in 1982, the famed "Bambi Bucket" appeared. Invented by Don Arney of Delta, B.C., it had rubberized fabric with aluminum tubing to provide structure, and with a 320-gallon ability, it was collapsible. What distinguished it from all other buckets was that pilots could use a hand controller to open and close a valve in the bottom of a Bambi, allowing them to drop water in multiple locations with the same load. As they burn off fuel, they can take more water, increasing their flexibility. How did it get its name? Some believe that it was named after the deer in the classic Disney movie. The more imaginative think that it honours a waitress named Bambi who worked in a pub in Boise, Idaho, where forest firefighters hung out. Neither is true. Arney claims that he had simply decided to call it "Bambi."

The changing needs of Conair's clients in 1978 suggested that helicopters would be a good addition to the company fleet to complement their

With 90 percent of the market share of buckets worldwide, the "Bambi Bucket" was invented by Don Arney of Delta, B.C., in 1982.

fixed-wing capabilities. Frontier Helicopters at Watson Lake was bought and moved to Abbottsford with its six Bell 47s and 206s.

• • •

Terry Dixon had flown for Okanagan Helicopters, doing heli-logging, firefighting, heli-skiing, and construction. He was made general manager of Conair's helicopter division.[6] He explained:

> One of the main advantages of helicopters was that they allowed us to deploy rapattack teams. Rapattack crews are trained to rappel from a helicopter into areas that are hard to access by foot or by vehicle and where there are no suitable landing areas for helicopters nearby. Three to six crew

Helicopters are essential in aerial firefighting, delivering rapattack teams, water from belly tanks, some with rapid-fill snorkels for refilling, and aerial ignition.

> members would rappel down to a point near the blaze, clear a helipad, and act as an initial attack team.
>
> Ideally, we would use them for small fires that had just started, usually the result of lightning strikes. If it was a small blaze, they could wrap it up without drawing on ground crews. If the fire was bigger, the crew would rappel down and act as an initial attack team so helicopters could later bring in firefighting crews. The warhorse helicopter was the Bell 205, the civilian equivalent of the famous UH-1H Iroquois Huey.

Although omnipresent in fighting wildfires today, the roles of various helicopters and their equipment are little understood, the media's attention going to their water scooper and retardant bomber cousins.

Helicopter Types

There are four size categories of helicopter for forest fire work in Canada, all now turbine powered.

Light Helicopters

Flying a pilot and up to three passengers, they are rarely utilized in fires due to their limited capability. Typically, they are for reconnaissance, mapping, and infrared scanning. Examples are the Bell 206B, MD500, EC120B, and Robinson R66.

Intermediate Helicopters

They can carry a pilot and four to seven passengers. They are the most utilized by forest fire management agencies for initial attack, crew transport, and general fire line support. They can do bucketing, crew moves, and equipment transport, as well as aerial ignition. Examples are the Bell Long Rangers, Bell 407, Airbus 350 series, Agusta 119, and MD900.

Medium Helicopters

They carry a pilot and up to eight passengers. They are able to move crews, equipment, and sling fuel, as well as bucket a fire; they are the "workhorse" of fire suppression operations. Examples are the Bell 204B, Bell 205A-1, Bell 212, and Bell 412.

Heavy Helicopters

These can carry more than twelve passengers and have a slinging capability of more than seven thousand pounds. Used for aerial construction and logging, few are available for casual hire. Examples are the Sikorsky S61, Boeing Vertol 107, Kamov KA32, Aerospatiale AS332, Boeing 234, and Sikorsky S64E.

Tanks

The belly tank has an external suction hose to allow refilling from remote water sources. Although heavier than a bucket, the tank system allows a faster cruising speed and is used near urban areas where sling loads are prohibited, due to the danger of the load being inadvertently dropped. Water can be dropped from a hovering helicopter to refill ground relay tanks if required. Helicopters with belly tanks require a larger area because they must hover at a lower height as their snorkels are usually only ten feet long. The water source needs to be free of branches, muskeg, and weeds, which can damage the bucket or door mechanisms. Rapid-fill snorkels allow for refilling in minutes and not returning to base. Nor do they require more than a couple of feet of water depth, but with the intense suction, they are susceptible to obstructions from swamp debris, muskeg, and leaves. Conair developed the Self-Loading and Off-Loading System in cooperation with the BCFS for use with the Bell 205/Belly Tank Helitanker. The system permits self-loading through the automatic deployment of a probe loader (suction hose and submersible pump) and offloading to a ground reservoir.

Dip tanks are portable reservoirs trucked in and set up where there is road access. These steel tanks can be up to forty feet long and ten feet deep.

On larger fires, where possible, these tanks can be set up close to the fire, which can shorten the turnaround times for the helicopters.

Aerial Ignition

Especially seen in the Donnie Creek Wildfire, planned aerial ignitions remove forest fuels and bring the fire's edge to established control lines. Using a heli-torch, this involves dropping a fire source away from the main fire to create a preburned fire guard.

Originally, aerial ignition was done by gasoline mixed with a gel (napalm) in a forty-five-gallon drum suspended under a helicopter. The amount is pumped and controlled from the helicopter with ignition provided either by a propane torch or electronic ignitor attached to the drum.

The Red Dragon developed by SEI Industries in Delta, B.C., is a helicopter-mounted device for controlled burns. The company's Sling Dragon dispenses one-inch Dragon Eggs injected with flammable ethylene glycol from a hopper slung beneath a helicopter, delivering up to 225 per minute and utilizing available GPS tracking.

Tactics/Roles

There are several phases of helicopter usage in forest firefighting.

Direct attack involves dropping water directly on the fire to cool it and reduce intensity. A heavy helicopter like the CH-47 Chinook, which can make heavy drops with precision, is used for direct attack with a water bucket or fixed belly tanks. These machines are rarely for crew transport or regular slinging support work. In most cases, where permitted, helicopters will utilize foam, which is added either to the bucket or tank for direct attack. Foam breaks the surface tension of water, allowing it to penetrate the forest more effectively. It also lasts longer before evaporating.

Medium and heavy helicopters can also drop fire retardant from a dip tank using the standard bucket or belly tank if required.

Initial attacks involve a small firefighting crew and equipment dispatched to a fire immediately after discovery. This is the most efficient and effective method of fire control.

If the fire grows, more helicopters are required to transport more crews and support them. Typical flying on larger fires includes reconnaissance, crew moves, equipment transport, and fuel replenishment for construction equipment that may be working on the incident. One helicopter will be identified as a medical evacuation machine should there be a need.

On the larger fires with more helicopters involved, an HLCO (Helicopter Coordinator) will be assigned to supervise a group of machines actioning the incident. This is an experienced firefighter who directs the various machines bucketing on a fire, providing safety for all involved in a smoky, radio-chatter-filled environment. All helicopters hired by fire agencies in Canada are required to have satellite tracking systems that give a location report every minute. Most agencies have a dispatch system that provides verbal contact between the aircraft and a central location. Due to safety concerns, aerial firefighting operations before helicopters were limited to daytime. But in night-vision goggles, with thermal imaging and enhanced visibility systems, helicopter crews increasingly now combat wildfires at night. This has unique advantages: there are fewer aircraft, the winds usually die down, humidity increases, and fire visibility is improved.

The Fort McMurray, Alberta, fire in May 2016 demonstrated rotary wing effectiveness, where as many as 147 helicopters joined the 16 airtankers, the 2,794 firefighters, and 233 pieces of heavy equipment to battle what became a "Campaign Fire."[7] Multiple heli-teams were employed, new bases were set up, and dedicated air traffic controllers worked on the ground to ensure the organization and safety of all aircraft involved. Apart from helicopters engaged in the direct attack and firefighter transport, their biggest role was the logistical support of the Caterpillar heavy equipment working on containment lines. As these were working 24/7, they required helicopters for crew changes and parts, as well as moving in seemingly endless barrels of diesel fuel.

Lessons in helicopter use after 2016 were applied to future fires. The number of heavy-lift helicopters for firefighting grew in tandem with the hotter summers. This was also because the U.S. military began selling off Boeing CH-46 Sea Knights and CH-47D Chinooks, as well as Sikorsky UH-60A Black Hawks. This helicopter debuted to movie audiences in the 2001 film *Black Hawk Down*.

In July 2024, a Black Hawk with Airborne Energy Solutions starred in battling the wildfires in Jasper. Dave Canavan, chief rotary-wing pilot for Airborne Energy Solutions, said, "We're pulling out of the lakes, beaver dams, rivers — whatever we can find as the closest water source — and trying to slow the fire down."[8] The Black Hawk requires two pilots at one time, and the third rotates in on fuel stops every two hours. Canavan and fellow pilots Lucas Myers and Cody Walker worked eight and a half hours continuously, dropping up to 2,500 litres of water each time from a bucket.

• • •

The rapid evolution of helicopters in a comparatively short time has inevitably led to Unmanned Aerial Vehicles (UAVs), either fixed-wing or rotor. Safer and cheaper than helicopters, drones have replaced them in the military, media, search-and-rescue, and utilities industries. Increasingly used to map the fire's spread, identify hot spots, and track the movement of smoke and ash, drones can now provide real-time images, allowing firefighters to quickly assess a fire. Just as helicopters did, drones, especially the quadcopter, which can deliver supplies and retardant, will change the way fires are fought.

Single-Engine Air Tractors

The challenge in 2025 is that the airtanker fleet is not keeping pace with the mega-fire activity. The last CL-415 was built in 2015, and De Havilland Canada is still in the early planning and production phase for the CL-515. Ex-military aircraft and former airliners had to be carefully and expensively modified to carry retardant. Enter the "ninjas," as one pilot calls them.

The biography of Leland Snow rivals that of aeronautical innovators like William T. Piper and Clyde Cessna. At a time when the standard crop-dusting planes were wartime Boeing Stearmans, in 1953, Snow designed the S-1, an agricultural aircraft with increased payload and performance. Beginning production facilities in Olney, Texas, he founded Air Tractor in 1972, building his first turbine model, the AT-302, in 1977 and expanding with the AT-802 in 1991. Initially sold as crop-duster aircraft, Snow's Air Tractors were adapted to narcotic crop eradication, cleaning up oil spills, light attack reconnaissance, and, most importantly, dropping fire retardant.

As early as 1995, Conair had recognized the potential of the AT-802 Air Tractor Fire Fighter. To evaluate the aircraft, the company sent a delegation to Olney, and Tim Garrish was assigned the task of evaluation and dropping various loads. "Prior to taking a single seat 802 for my first flight," Tim recounted to me, "Leland took me aside and said, 'There are two ways to take off with the 802, Tim. On the runway or through the grass to the side.' If the tail wheel lock was not engaged, one would end up off the runway in the grass, as Leland had observed a number of times."[9]

Conair ordered two AT-802s for delivery in early 1996 and became the Canadian dealer for the AT-802F. British Columbia replaced their Conair Firecats with two AT-802s, flown by Tom Zurowski and Tim Garrish, basing them in Williams Lake for the 1996 fire season. In 2024, Conair operated the largest fleet of AT802 (seven wheeled and sixteen Fire Boss amphibious) in the world.

Three experienced air tractor pilots explained the use of the aircraft to me.

Guy Cannon

As a boy, Guy Cannon used to fly with his father in the family Super Cub, often getting airsick but always enthusiastic to go again. He managed to take the controls as soon as he was big enough to reach them.

Years later, he found himself at the Wipaire facility on Flemming Field, South St. Paul, Minnesota. There, suspended over a set of floats, was the first Fire Boss bomber in its early design stages. Bob Wiplinger, WipAir's patriarch, told Guy, "One day you'll be flying one of these."

Guy elucidated:

The basic Air Tractor, model AT-802, is a taildragger, where the pilot sits alone in the middle, controlling it with a joystick, much like a Second World War ground attack fighter. Modern technology has improved function and reliability, allowing the 802 to carry the payload of a wartime bomber to targets with precision. Leland Snow believed it would be too big and too fast for agricultural work and that they'd be lucky to sell a dozen of them. (I recently flew serial #1069.) It has a roll cage to protect the pilot, a safety glass windscreen, reinforced wing leading edges, a skid plate in the roof, airbags built into the five-point safety belt harness, emergency escape doors, and even a cable connecting the top of the canopy to the top of the rudder in case the airplane flies through the hydro lines, to keep from shearing the tail off. Also, the pilot must wear a helmet. Did I say robust?

In September 2001, I was in Minnesota to pick up a friend's recently purchased small biplane. The airspace was closed due to the horrific events of 9/11 in New York, so we drove down from Thunder Bay, Ontario, with a trailer to transport the plane back to Canada. Looking for some aircraft-suitable tie-down equipment, we stopped at Wipaire at Flemming Field to see an old friend, Bob Wiplinger, and ask if he had any suggestions. As was always Bob's nature, he was happy to help and invited us to his weekly hangar BBQ on the Mississippi River. He said to me, "Hey, if you've got a couple minutes, I've got something you'd like to see." Of course, one never passes up such an opportunity, so we went over to the hangar adjacent to his office to see the latest project, the Fire Boss. At this time, it was an AT-802, looking a little worse for wear, along with a collection of structures and drawings that would become the first of many.

Bob explained they were taking their 13000 model floats for the DHC Twin Otter and reengineering them

Guy Cannon with Fire Boss C-FFNT on July 31, 2018, in North Bay, Ontario. After many seasons of aerial firefighting, he observed, "People are usually pretty happy to see you."

for the Fire Boss. Although the Fire Boss would outweigh a Twin Otter by nearly two tons, the floats could be smaller, and thus the aircraft was more efficient. The Fire Boss would not be designed to land with a load. It would sink if it did! Much modification and test flying then ensued.

My eventual instructor, Mark Matheson, did the pioneering flying with the newly designed floats, stability mods on the tail, vortices generators (to help lift on the wings and tail), bigger propeller, and bigger engine, taking the project through certification by 2013. The first Fire Bosses were single-seaters. To pass on his wisdom, Mark "cheered on" new aviators from a boat with a handheld radio while the pilot figured out the aircraft with his guidance. Eventually, a two-seater was built, which I flew with Mark, along with many others. At that time, I was also an instructor on the 802, but on wheels, and Mark and I lamented the similarities of our professions. "We sit in the back and say bad words while the newbie in the front tries to kill us." We laughed. Our job, simply put, was to try to prevent that from happening.

The Air Tractor Fire Boss is basically a flying tank with a propeller on the end, without all the extra space for passengers and luggage of some of the aircraft converted from other uses. This makes them very efficient and effective. The aircraft weighs sixteen thousand pounds loaded and can carry more than its own weight in fuel (1,400 litres) and water (3,200 litres) when full. There is a foam injection system that allows the water load to become a sticky mess that hangs on to vegetation. The amphibious floats equipped with retractable wheels allow the aircraft to be reload with retardant at airfields.

When scooping from water, a clear run of a mile or so is sufficient without obstacles. The water run takes roughly thirty seconds, about half of that spent with the scoops down, picking up the water load.

One of the great features of the plane is the fire gate. Or, in aviation-speak, Fire Response Dispersal System (FRDS). This computerized system allows the pilot to choose from a wide variety of options for the drop. A salvo load opens the gate doors completely, allowing the contents to be dropped immediately. This is effective for tall trees and heavy canopies, where the force of the load is needed for the drop to make it through to the ground. A dialled-back coverage setting allows the pilot to string out the drop in a longer line. This is very effective for grass fires and such, since there are numerous settings. A partial drop, or split load, allows the pilot to hit multiple targets on the same run. Drops are done from a specific height and speed to be most effective. Too high or fast, the drop dissipates, having minimal effect when it hits the ground, if it even gets there. Too low, there's a risk of breaking treetops and branches, as well as "shadowing the target." This is where the load, still in forward motion, only coats one side of the target, allowing the fire to find the uncoated shadow of the drop and continue burning. Just right, about a wingspan above (sixty to a hundred feet) and 105 knots allows the drop to effectively coat the target vertically, stopping the fire's progress.

Most effective when used in groups, the Fire Boss is an initial attack medium bomber to get at the fires quickly while they're still small. If a suitable water source is near, Fire Bosses have a significant impact. Our group of four's best showing so far is 250 drops in two and half hours. The ability to stay on the fire and not allow it to spread between drops is a crucial factor in preventing small fires from becoming big fires. The water scoopers excel at this, and the Fire Boss is an effective tool to add to the arsenal of the fire crews. It is most satisfying to stay on the fire, watching the suppression efforts take effect in real time.

Supporting the ground firefighters, knocking down the smoke and flames, protecting and saving values, is most gratifying. A successful mission brings about a true sense of achievement. A Fire Boss captain is always looking forward to the next mission and doing it all over again. Preferably the next day.

• • •

After many seasons of aerial firefighting, Guy Cannon observed, "People are usually pretty happy to see you." When asked about the rewards of the job, he says, "The satisfaction of a job well done, resources protected, and crisis averted are more than enough reward to make the career choice the best flying job I could ever get."

Brad Blois

Brad Blois explained his passion:

> Your goal, no matter what the alert level or expectation, is to be safely in the air as soon as possible. Fires don't wait, and minutes count in the early stages of a forest fire. Nobody does anyone any favours by rushing. In extreme hazard conditions we spend all day at the airport close to our airplanes. My personal preference is to hang a hammock between the floats of my Fire Boss, and I wait in the shade of the airplane.
>
> In all cases, you arrive in the morning to DI (daily inspect) your airplane so that it is ready when the call comes. This involves an inspection and test of all the systems. Most guys will do their checklists to the point of engine start and leave the airplane ready to start and go at a moment's notice. Some jurisdictions do a morning briefing, while others send it out digitally, outlining the fires, weather, and all associated activities and risks in the area.
>
> Then there's the waiting, which is basically what a fire bomber pilot's job is most of the time. You wait for a call. Most times it never comes, but you must always be ready for it. This is the stress of the job, and one of the challenges as well. Staying alert and fresh is a big challenge during the long summer days. Some jurisdictions do a weekly practice, which involves a flight and a few simulated fire drops. For those that do, it's wonderful and appreciated by the pilots. For those that don't, you need to stay fresh mentally. Everyone has their own way to do this; it requires self-discipline. My routine is: After I do my DI and my checklist, I select a different emergency procedure each morning to review. I verbalise and simulate each item so they are fresh in my mind and my muscle memory for that rare event that I may need them.

Now for the moment every aerial firefighter lives for: dispatch! Honestly, half the time we are given a warning before the phone call happens. This is great because it allows you to prepare yourself physically and mentally for what's coming. If you have enough time, you dial up the fire area on your device and start reviewing the risks, terrain, hazards, and water sources if you're flying an aircraft that reloads in the water. When the call comes, it's an organized scramble to put on your safety equipment, which, depending on what you're flying, consists of fire-retardant coveralls, life jacket, helmet, and HEED (Helicopter Emergency Egress Device). It's important not to rush. Looking back on my career, all the biggest mistakes I have made were when I rushed. But we don't waste any time, and engines are usually running in minutes, coordinates to the fire are shared, air traffic is notified, and we are given priority to depart ASAP.

The fire can be anything from a grass fire or a smoking tree to a brewing conflagration of all kinds of fuels. After takeoff and our safety checks (calls made to control, group members, or birddog) we either fly directly to the fire (if we loaded at the airport with water or retardant) or fly to the nearest source of useable water.

The first thing we see is the smoke column on the horizon. Even a small fire usually throws up enough smoke to spot it from miles away. If it's a big fire, everyone's hearts are beating pretty fast now, and adrenaline is flowing. At least for me, that is, as I prepare myself for the unknown and start to formulate a plan. That plan consists of the best routes to and from the water, the safe areas, what my hazards are, and, very importantly, what other traffic is in the area. I note wind, areas of turbulence, and start to observe the fire behaviour. Meanwhile, as you approach the fire the radio is buzzing with chatter. Instructions and calls flow continuously on the fire channel, and you need to keep a continuous focus on it to maintain situational awareness in a constantly changing environment. If you think that sounds tough, the birddog pilot has it even worse, since they are talking to everyone, plus all the guys and gals on the ground, from the firefighters to the heavy-equipment operators. If you've ever seen a disturbed hornets' nest, well, that's the airspace around an active forest fire.

Next thing that a pilot notes is, surprisingly enough, the smell of the smoke. It carries a long way from the fire, and it smells very familiar to a fire pilot. In fact, I have smelled a fire a few times before I have seen it and we've gotten the call to fly.

It can simply be described as a corruption of that pleasant smell when you have a campfire in your backyard, camping, or on the beach, but magnified greatly, and with a menace to it. You get used to the smell, which can become a stench after hours of flying over it, in it, and through it. If communities burn, that's even worse, and it is a stinking, poisonous smoke to everyone involved when all the various things in a person's life go up in flames. Fortunately, that situation is comparatively rare for us.

In addition to the smell of the smoke, its colour tells an experienced fire pilot a lot about what is going on. A grass fire throws a familiar light-coloured smoke, but when a fire gets into heavier fuels like forests, it turns black, and the more it burns, the higher it climbs into the sky. Man-made structures when burning make a terrible black flame that seems angry as it burns hot, cold, and dances into the air, depending on what is burning. Surprisingly enough (or not, considering the fuel, plastic, fabric, and rubber), vehicles burn quite readily and sometimes are a contributor to starting wildfires; their black clouds of smoke are easily spotted from afar.

That brings us to the actual fire, of which there are two kinds, with the second having two categories that we face as part of our job. They are grass/brush fires and forest fires. The latter are divided into two groups, new starts and campaign fires.

Grass/brush fires are exactly that. They start in the grass and bushes, in most cases by accidental human activities, such as farm equipment, a garbage fire that got away, hot exhaust on an ATV or motorcycle, or even a cigarette on the side of the road. Grass fires are common, travel very fast, and can frequently get away from ground forces. They are not without risks, however, as they are usually near built-up areas and frequently feature dangers to pilots, such as power lines, poles, and towers.

Forest fires are our main job. They are dangerous, destructive, and full of potential for disaster. Our limited resources are best focused on initial attack: hit the fire hard and fast with everything you have before it becomes a big fire. For a pilot flying above or around it, there are several key things to focus on and a few that we can do to help the guys on the ground.

One is for water bombers to throw a few loads into the head of the fire to knock some energy out of it so the smoke dissipates and everyone can get back to work. It's not uncommon for air assets to be grounded from smoke despite the desperate need for them at the fire line.

This is where initial attack is key. On smaller fires, we can attack the head of the fire, where we can do the most effective work. Once the fire gets too big, the head becomes too dangerous to work. It starts burning too fast, creating very dangerous conditions for aircraft and ground crew. As cold air gets sucked into the fire, the flames and smoke rise rapidly into the air, creating massive turbulence and unsafe flight conditions. Often these hazards are invisible, so it's always difficult to judge if it is safe to attack the head or not. My best friend was killed by an invisible fire tornado without warning that there was even a hazard on what was a normal fire. But when conditions are right on smaller to medium-sized fires, there's nothing more satisfying than putting loads of water on the head of a fire and seeing real results. On these days, I love my job.

Once a fire gets into heavy fuel and bigger trees, it creates a few problems. The canopy shields the fire from our drops, so increasing the rate of the drop (usually to maximum) punches through the canopy and gets the water to the ground. But when burning trees start to "crown" (fire rises rapidly up the tree and flames engulf the top, like a crown), they throw burning ashes up into the air, which are blown downwind, starting more fires ahead of the main fire. So, crowning trees and the hot spots ahead of the fire line are frequently high-priority targets for water bombers.

However, sometimes luck isn't on your side ... maybe the wind picked up, the temperature crossed over (crossover is the point where the temperature exceeds the dew point, and it makes for dangerous fire conditions), maybe there's not enough ground crew available, or the fire itself is inaccessible. When a forest fire gets big and stopping it is no longer a reality, then it becomes a campaign fire.

These are exhausting, frustrating, and dangerous. Exhausting because you're battling the worst Mother Nature can throw at you for days on end. Frustrating because you seldom see good results and are left watching things be destroyed at the whim of the fire under your wings. Dangerous because you're spending hours every day low to the ground in the worst heat, smoke, and turbulence you'll ever encounter in an aircraft. Your harness needs to be strapped down tight, and there's no time for rests or breaks during your fuel cycle. It's just

> scoop, drop, return, for three hours and then home for fuel, a snack, and some water, and repeat, repeat, repeat. Fire Bosses are very physical aircraft to fly, and it's a real workout for your limbs, especially pulling full loads of water out of the lakes in hot weather. You get very used to your stall horn sounding and your helmeted head hitting the sides of the airplane.

Juliana Turchetti

Juliana Turchetti enthused:

> Today, on the 14th of February, 2024, at 10:11, I did my first solo flight in the Fire Boss. I am grateful to everyone and everything that led me to this moment.
>
> About a week ago, I had my first scoop lesson, and after landing, I had this tickling sensation in my heart and my stomach, very, very similar to what we feel when we fall in love.
>
> Airplanes are just machines when seated at a ramp — and humans are just humans when walking the Earth, but when these two are put together, it's an explosion of magic, aerodynamic passion … a fusion of techniques and feelings, a symbiosis of our desire for flying and the airplane necessity of a human to fire it up.
>
> And we became one. A poetic unity of man and machine!
>
> There's something special about Air Tractor Fire Boss single-seat aircraft. Its nature is challenging and inviting at the same time. The Air Tractor is more than an airplane — it is a statement. When boarding, the pilot needs to climb on the floats and then make it through two sets of steps to reach the wings and then the cockpit. Just getting aboard already provides an upper view of the world. The magnificence of the Fire Boss is undeniable. It doesn't matter if it

Juliana Turchetti: "Me and my Valentine! Today, on the 14th of February, 2024, at 10:11, I did my first solo flight in the Fire Boss.... I had this tickling sensation in my heart and my stomach, very, very similar to what we feel when we fall in love."

is at the ramp or flying, if you want to see it properly, you will have to look up.

In February 2024, as a newly hired pilot, I was given the introductory flight, and it was thrilling! A wonderful combination of adrenaline, dragonflies in my stomach, and the desire of being able to tame that bull. From the first day, we practiced manoeuvres, such as turns, stalls, takeoffs, landings, scooping, and water droppings. Appropriately, it was Valentine's Day when I first soloed that incredible machine.

And I fell in love with it. The sensation of achievement, freedom, and power felt incredible! The whole training is also a humbling experience. At the beginning, when learning the art of scooping, we endured a lot of porpoising.

> There's a reason why the scooping button is also called "the rodeo button." That bull will definitely show its power! We must learn the balance between riding along and taming that animal. With time and practice, things start making more sense. It still amazes me when I think about the process and the purpose of putting this airplane on the water!
>
> The development of intimacy and trust between pilot and airplane takes time — as in in any romantic relationship — but must be pursued in every interaction. Single-seat airplanes normally require multitasking abilities. We work as a team, and we struggle individually. It's an eternal waltz where you make yourself present but still think ahead and are planning your next step. And then, hopefully, it will become second nature. I am looking forward to my first fire season. It will be challenging, no doubt. Air Tractor and Fire Boss — the environment is very dynamic, and we rely not just on our own abilities and the airplane capabilities but also on our teammates, on the ground and in the air. And they rely on us.

• • •

Juliana was killed on July 10, 2024, working to control the Horse Gulch fire in Helena, Montana.[10] Her employer, Dauntless Air, spoke for all who knew her.

> Juliana Turchetti was an incredible aviator. More important than that, she was an incredible person. The outpouring of love and joyful memories we have received for Juliana over the past twenty-four hours are a testament to the impact she made on people around the world. She blazed trails as one of Brazil's first female agricultural pilots. Every person fortunate enough to come across

> Juliana was embraced with her enthusiasm, kindness, and care — whether she knew you for a moment or for a lifetime. You could feel her passion for flying right away; it was infectious. Her belief in our mission to protect people and communities was unwavering. She had a commitment, determination, and energy that inspired us all. Blue skies and tailwinds to you, Juliana. Forever part of our family.

• • •

We are living in what climate scientists call the new age of exurban firestorms, where homes and rural industries are increasingly built in areas of high and growing wildfire risk. Requiring less infrastructure than large tankers, helicopters and SEATs are quicker to load, more precise in discharge, and the former can even operate at night. Innovation has always been part of aerial firefighting, and with such wildfires growing in number, size, and intensity, companies know they must quickly adapt to both platforms.

EPILOGUE

The Gods Didn't Warn Us

AS CANADIANS, WE IDENTIFY WITH WINTER AND SNOW BUT DON'T REALIZE that fire is the defining element in our land — and our lives. Our forests and prairies evolved through fire from the last Ice Age. But as poet Lorna Crozier, watching the fires that devastated British Columbia, penned, "The gods didn't warn us / fire has no heart."[1]

When it comes to aerial fire suppression, Canada is a superpower. Years before the United States, the Ontario Provincial Air Service initiated wildfire detection from the air. One of its pilots, Charles "Carl" Crossley, invented the world's first effective water-dropping system. Our Canadair water bombers are so sought-after that half a century after the first one flew, they are being rebuilt in modern form, and there is a twenty-year waiting period for them. After the tanker cancellations in the United States in 2004, both Transport Canada and Conair were held up as examples by the U.S. government of what the Forest Service and U.S. tanker companies should be.

For much of the twentieth century, suppression was the sole means of fire management. The more we suppressed fires to protect communities, the harder it became to protect those communities; dense undergrowth is the tinderbox and climate change the spark.

Instead of focusing on putting out flames, wildfire agencies and provincial governments must carry out fundamental changes to prevent fires from igniting and spreading in the first place. Proactive programs such as reducing severe wildfire risk by managing forests ecologically are beginning to be implemented. Ecological forestry includes practices like strategic thinning of forests, prescribed burning by igniting small, controlled burns in targeted areas to reduce undergrowth, and managed wildfire. Managed wildfires are non-planned fires that are allowed to burn without being extinguished under certain circumstances. These practices mimic the way that Indigenous Peoples historically managed forests before colonization.

But after a century of being indoctrinated that fire is the enemy and its only mission is to destroy civilization, the public is justifiably afraid of prescribed burns and managed fires. That they might cause larger, out-of-control wildfires is an aversion that stems from that fear. That fires are necessary to maintain the health of the nation's forests is difficult to grasp.

Today, mega-fires in towering columns of heat are detected from satellite platforms. Satellite imagery and mapping tools provide an overall picture of fire activity to identify the general location, extent, and intensity of wildfire activity and its effects. In 2019, the California Department of Forestry and Fire Protection, known as Cal Fire, gained access to a systematic detection tool, an artificial intelligence (AI) program. Harnessing AI and deploying it to prevent small fires from becoming conflagrations was greeted with the enthusiasm that the introduction of aircraft once did. The artificial intelligence software was able to alert firefighters of the presence of smoke even before dispatch centres received 911 calls. The fires that were still small and manageable were extinguished. But the program's apparent success comes with caveats. The system can only detect fires visible to the cameras. And humans are still needed to make sure the AI program is properly identifying smoke.

The United States military offered to use its highly classified network of spy satellites as well as drones and other aircraft to alert state agencies when fires were detected. Unfortunately, the orbits of existing satellites, including NASA's Visible Infrared Imaging Radiometer Suite (VIIRS) satellite and its Moderate Resolution Imaging Spectroradiometer (MODIS) satellite, do not take them over Canada during the times most crucial to firefighting efforts.

The "peak burn period" is between about 3:00 p.m. and 6:00 p.m., when wildfires tend to be at their most active. Higher temperatures, stronger winds, and low humidity in that period contribute to the rapid spread of fires.

Canada is preparing to plug that gap in 2029 by launching the $170 million WildFireSatsystem. Its three satellites will be equipped with infrared sensors that will detect and measure the "fire radiative power" emitted by wildfires. This measurement means wildfire services can calculate fire intensity and rate of spread, in real-time, precisely when fires are most dangerous. The WildFireSat initiative is to directly involve Indigenous Peoples and nations, elders, and knowledge carriers, groups, and organizations in fire and emergency management. It will make opportunities for Indigenous-specific initiatives for Indigenous land, fire, emergency, and social services managers and groups to participate in culturally informed resources related to the capacity-building activities. In 2023, the wildfire crisis disproportionately affected Indigenous populations nationwide — as have all previous ones. It highlighted the importance of engaging with Indigenous communities and respecting their traditional knowledge.

• • •

Our forests make up nearly 9 percent of the world's total forest area, and throughout this century, we will continue to see larger, more frequent, and more destructive fires. You can't fireproof the forest, and there is no single solution to fire. But by 2030, we will have some difficult decisions to make. Should we be hiring more firefighters from abroad? Building hundreds more water bombers? "They can't stop these fires," claims Steven Pyne, the environmental author. "I mean, they could have fifty thousand firefighters there now, and it's not going to change it. We could have two hundred more air tankers. Are they going to be able to stop these fires that are going? No."[2]

As North Americans, we have difficult choices to make about how to confront wildfire. Aircraft, UAVs, drones, AI, and sensors mounted on trees are stopgap measures until the next fire season. Ecological forestry — and eventually zero-emission industries and vehicles — are beginnings in turning back the tide and avoiding what is otherwise sure to be a dystopian future.

Acknowledgements

WITHOUT WORDS, WITHOUT WRITING, AND WITHOUT BOOKS, THERE WOULD be no history and no concept of humanity. This is certainly true of aerial fire suppression over the last century. The sources I have drawn on for *Fire Eaters* are many and varied. In addition to the large number of books and websites consulted, many people have provided personal accounts to this history, and for that I am most grateful. Mariah Leuschen-Lonergan, public affairs specialist, U.S. Forest Service, found me Emma Hawn, Aaron Thorp, and Jonathan Fuentes, three smokejumpers to interview, and I thank her profusely. Pierrick Caillet, technical manager of the Canadair CL-415 fleet, French Sécurité Civile, earned my gratitude for his unfailing assistance in locating historic photographs of the Canadairs.

"Writing is a lonely job. Having someone who believes in you makes a lot of difference. They don't have to makes speeches. Just believing is usually enough."[1] I owe incalculable thanks — and a trip to Paris one day — to Donna Hudson, my wife, to whom I expound daily what I have written — without mercy. After my last book, *American Obsession: Howard Hughes and Juane Trippe*, she begged that there would be no more stories of Hughes's sordidness at the dinner table. This time, I could tell her about H-Boats and Catalinas, explain "fracture fatigue" and the Wildland Urban Interface, and wax lyrical about birddogs and unappreciated smokejumpers. Having grown up in fire-prevalent British Columbia, she lobbied to have the book's subtitle be "Saving the Forests from the Sky." As Richard Dreyfuss repeatedly said in the movie *Always*, that's my girl.

Writing non-fiction is like engaging in organized fraud. You are laundering the experiences of those whose lives are far more interesting than yours into readable currency. My childhood at the edge of Bombay's Santa Cruz Airport (now Mumbai and Shivaji International Airport) and my adult life representing the Canadian government in Hong Kong, New York, The Hague, and Vienna (and much later as a journalist in places like Kandahar and Khartoum) were all far from forests. What I knew was a drop; what I didn't know was an ocean. In writing *Fire Eaters*, I was incredibly fortunate to have met intelligent, generous (with their thoughts and time), risk-taking people, and I appreciate them for what they could teach me. They wrote of the intricacies of flying and maintaining airtankers, helicopters, and water scoopers, some even while fighting fires in Australia and Texas. Boundless gratitude goes to Guy Cannon, Ray Horton, Tim Garrish, Terry Dixon, Megan Nunner, Ray Bozzer, Brad Blois, the late Juliana Turchetti, and the men and women at Neptune Aviation. All of you found something worth defending in your lives, and your insights and enthusiasm explain what I only knew from television and YouTube. Thank you to everyone who has helped bring this story to light!

When I asked Juliana Turchetti to write something for the book, her immediate response was "It's almost like the words are begging to go out and let people know how it feels to experience what I do when I'm up there." I think of her with this quote: "She cried like a child, laughed like bells ringing, and her smile was the most beautiful I've ever seen. She loved, she hated, and she inhaled the world like it was a rose. All this was why I loved her more than life."[2]

As always, since this is the thirteenth book of mine that they have published, I owe a great debt to Dundurn Press. Thank you, Meghan Macdonald, Kathryn Lane, Elena Radic, Laura Boyle, Dominic Farrell, and Russell Smith, for having faith in me.

For the history of aerial fire suppression in both the United States and Canada, I chose two companies that I thought best defined the characteristics of each country: for Canada, the Conair Group Inc. of Abbottsford, B.C., was an obvious choice. President and CEO Matt Bradley believed in the book from the start. Shannon De Wit, Conair's public relations and communications manager, was generous with her time and contacts. Somehow, Neptune Aviation Services' Jennifer Draughon and Jennifer Cles, gearing up for another wildfire season, found the time to collect their employees "sea of memories," as one pilot wrote. Again and again, they enthused, "Battling fires is dangerous, but I love my job, and it's because of the people at Neptune — I love our family dynamic." Your words took us readers away and brought us back better made. We don't naturally fall into perfect relationships. We create them.

Finally, thanks are due to good friends John and Carol McKay, John and Joyce Clapp, and Nancy and Ted Lennox, all of whom through the years have provided me with endless good humour, stimulating conversation, and gin and tonics, in that order.

• • •

The final draft of *Fire Eaters* was completed on April 26, 2024, exactly 124 years after the Great Fire of 1900 destroyed two-thirds of my hometown, Ottawa, incinerating all houses that were on my street. Warnings had been ignored for years about the increasing piles of combustible lumber and wooden buildings crowding both sides of the Ottawa River. The gods didn't warn us. Fire has no heart.

APPENDIX 1

The Curtiss HS-2L Flying Boat

Technical Information

Wingspan	22.6 m (74 ft. 19/32 in.)
Length	11.9 m (39 ft.)
Height	5.7 m (18 ft. 9 1/4 in.)
Weight, Empty	1,950 kg (4,300 lb.)
Weight, Gross	2,918 kg (6,432 lb.)
Cruising Speed	105 km/h (65 mph)
Max Speed	137 km/h (85 mph)
Rate of Climb	550 m (1,800 ft.)/10 min
Service Ceiling	2,800 m (9,200 ft.)
Range	830 km (517 mi.)
Power Plant	Packard Liberty, 360 hp, V-12 engine

APPENDIX 2

Vickers Vedette Technical Information

NAMED FOR FEMALE CABARET ARTISTS OF THE DAY, THE NIMBLE VICKERS Vedette made the HS-2L look like a blowsy dowager who had seen better days.

General Characteristics

Crew	3
Length	32 ft. 10 in (10.0 m)
Wingspan	42 ft. (12.8 m)
Height	11 ft. 9 in. (3.58 m)
Wing area	483 sq. ft. (44.9 m^2)
Empty weight	1,849 lb. (839 kg)
Gross weight	3,049 lb. (1,383 kg)
Fuel capacity	535 lb. (243 kg) fuel and oil

Powerplant	200 hp Wright J-4 Whirlwind engine or Armstrong Siddeley "Lynx" 215–230 hp engine
Propeller	2-bladed fixed pitch plusher propeller
Performance	Untested
Maximum speed	59 mph (95 km/h) at sea level; 87 mph (140 km/h) at 10,000 ft. (3,000 m)
Cruise speed	47 mph (75 km/h) at 5,000 ft. (1,500 m)
Endurance	5 hours
Service ceiling	13,000 ft. (4,000 m)
Rate of climb	650 ft./min (3.3 m/s)
Time to altitude	10,000 ft. (3000 m) in 30 minutes

APPENDIX 3

Neptune Aviation Services

TO REPRESENT THE UNITED STATES, I WANTED TO PROFILE A COMPANY THAT symbolized the underdog — the mutt of the pack, as one of the pilots called himself — the Jimmy Stewart (or Tommy Lee Jones) marshal who rides into town to protect the small rancher from the evil cattle barons. This is the culture that invented Hollywood, and I went to the movies for larger-than-life characters like Pete in *Always*, making a steep dive dousing Al's engine with slurry, Rocky Balboa running up the steps of the Philadelphia Art Museum, and — this might be a stretch — Pete "Maverick" Mitchell buzzing the tower. And that was Neptune Aviation Services, Missoula, Montana.

In Norman Maclean's *A River Runs Through It and Other Stories*, set in Missoula, Montana, fly fishing is an allegory for life. If you can "read" the river, the fish, and the whole world — if you can do what needs to be done, with honesty, resilience, and love of family — you will have attained the grace to live a good life.

For one airtanker company, reading the river meant never giving up, even when faced with crushing adversity. The decision in 2004 to cancel contracts for all heavy firefighting aircraft not only had a grievous impact on many states in their efforts to fight forest fires but also on small companies like Neptune Aviation Services in Missoula, Montana.

The indomitable Marta Timmons purchased Black Hills Aviation in 1994, with its fleet of Lockheed P2V Neptunes retrofitted for aerial firefighting. She brought it to Missoula, renaming it Neptune Air Services after the aircraft. When the abrupt cancellation of contracts occurred on May 10, Neptune Aviation already had two aircraft fighting wildland fire. It was overwhelming for a small company that later also suffered three P2V crashes with devastating loss of life. But as in a Zane Gray novel, for the men and women in Neptune, the phrase "riding for the brand" meant remaining firm and resolute, cherishing one's work family, and going the extra mile.

Aware of its contribution to firefighting, when Neptune retired its last P2V aircraft in 2017, it donated all remaining ones to museums in California, Oregon, Michigan, and, appropriately, the National Museum of Forest Service History, Missoula.

Jennifer Draughon, president of Neptune Aviation Services, with the three-thousand-gallon-capacity Next Generation airtanker, the BAe-146.

"When the Forest Service cancelled all the large airtanker contracts, it was a really hard lesson," Jennifer Draughon, the president of Neptune Aviation remembered. "Although I wasn't the president of the company at that time, I witnessed how it was handled, and part of the way Neptune operates and part of how I was mentored, was you work together, and you steer together from leadership down."

Brad Ruble, Captain and Thirty-Eight Year Firefighting Veteran

I flew as a P2V captain with Neptune Aviation from the very beginning. What I liked about flying the P2V was its power. If you knew how to fly it, you could get a good climb rate going. You could use the power on or off, and you could use that power over a fire, which gave the P2V advantages over other aerial firefighting aircraft in that era.

Being on tanker bases and collaborating with the teams fighting fires on the ground fostered a spirit of duty; we felt a moral obligation to get to the fire and make accurate retardant drops. We were there to help the ground team, and we wanted to do an excellent job.

My most memorable retardant drop for Neptune was a non-traditional flight. In April 1997, First Officer Carl Fisher, Crew Chief Sean Wilcox, and I were based in Brainerd, Minnesota. The Red River had flooded Grand Forks, North Dakota, and fires had broken out in the downtown area. Fire trucks were unable to get into the city because of the floods. The fire had spread and was threatening multiple buildings. We dropped on a building that had spot fires on its tar roof. The guys on the ground said the retardant did its job and saved the building. A picture of that retardant drop was featured in *National Geographic* magazine.

From its early years, Neptune Aviation always evolved and moved forward. Its arc has always been to modernize and build a better company. We were always trying to get better. We had the knowledge, the expertise, and we had the drive to improve. We had a vision of where Neptune needed to go, and that vision drives us today. When Neptune retired the last P2V aircraft in 2017, I retired as well.

Cliff Lynn, Director of Maintenance

In 1999, I was sweeping floors and thought of going to college to be a biologist. But after starting at Neptune, I went on my first airtanker drop, and that changed everything. By 2005, I was a crew chief on an airtanker, travelling the country for the next ten years on one of our P2V aircraft.

When you're in the airplane, you love flying with firefighters and supporting those on the ground. It has an effect on you. During my time on crew, there were many memorable fires, but one of the fires that stands out took place near Redding, California, when the city of Weed was burned. Our team dropped twelve loads of fire retardant on that fire. When the pilot came back to base, we were loading, and we needed to hit the fire house by house to protect each of them from destruction. Fortunately, nobody died in that fire, but it was emotional. I was on the ground, looking at the smoke on the hill, and you see all these powerful men just sitting there in silence, thinking about what they just experienced. They were protecting schools, homes, and other buildings — and they had to do it fast. This is the kind of fire that people remember because you see how powerful fire is, and you realize that there are times when you are completely helpless.

I haven't been on the road as a crew chief since 2015, wanting to spend more time with family. We also moved on from our old fleet of P2Vs to BAe 146 airtankers around that time. The BAe is a great plane, but I thought those P2Vs were something special. That plane had a soul, and when you worked on it, you were intimately involved with the plane. It was one of the main reasons I stayed working as part of the crew team as long as I did. When you took off in a P2V, everyone at Neptune knew it. They not only heard the takeoff — they felt it, and I loved that. In fact, I have a P2V tattooed all the way across my back.

In 2008–9 we lost six people, including Gene Walstrom, who was like a second father to my brother Nic, and Brian Buss, who was like a brother to me. A few years later, another close friend, Todd Tompkins, was killed in an airtanker crash, and within months I had a daughter who was stillborn, and then my wife passed away.

Jennifer Cles, Vice President of Finance

I joined Neptune in August 2013 with a background in construction accounting. I came on board as the controller, the same month we were doing grid testing for the BAe. It was a whole new and exciting world.

People tend to think that accounting is a boring profession, but that could not be further from the truth when the business is aviation. Every day brings surprises, and our industry is hard to predict. Projections and budgets are critical to our success, and yet it can be hard to forecast the fire season. We must balance staying flexible with planning ahead.

The people in Neptune like to play practical jokes, and I remember coming into the office one morning shortly before Halloween, unlocking my door, and seeing a horrible face staring at me from my office chair. It was the body of a skinny red devil!

Visitors aren't spared our team's pranks either. We have a test dummy with a monkey mask, yellow hard hat, and an orange flight suit perched up high on the storage bay, watching over the tanker maintenance hangar. I recently toured a class of children, and they spotted it and were quite intrigued.

Chris D'ardenne, Maintenance Control Manager

I always had a love for airplanes — especially old planes — Second World War vintage aircraft. I went to school outside of Philadelphia to secure my Airframe and Powerplant (AMP) license and a guy pointed out an ad for a job in Montana. I called, and they didn't have anything available at the time, but they called me back two months later, flew me out to Montana, and I joined the team.

During my first year here, back in 2000, I was sent out to Tanker 11, which was in Reno, Nevada, at the time. During winter heavy maintenance, we had installed an original navy mothballed engine onto our P2V-7 aircraft, which still had the original navy paint on the cowlings. I drove a truck and trailer from Missoula with the correctly painted cowlings and was hanging around the tanker base when we got a call to help with a fire.

This was all new to me; I had never seen planes fly over a fire before. The captain had called to Dispatch and asked if I wanted to go, and I responded, "Hell yes!" Just being on the plane for the takeoff was insane. The fire was burning right outside Reno, and after we dropped the retardant, I looked back and thought it was awesome. Fighting fire on the edge of town, flying this cool P2V — the whole experience was amazing, and I was hooked.

There was another fire in Idaho near the Sun Valley Ski Resort. We were dropping retardant way out in front of the fire, maybe two miles out. It was a huge fire, and we were circling it when we needed to drop retardant in a ravine. We needed to get close to the ground for the drop and were able to do it, but to get out we had to turn right down one canyon, then take a hard left down another. We had to follow two different valleys to get out, and it was a pretty wild experience.

A lot of people don't like being on the road, but I loved it. I liked the uncertainty of not knowing where I was going to be night to night. We used to move around a lot. You would wake up, pack your stuff, and you had no

The beloved Neptune P2V. Aware of its contribution to firefighting, when Neptune retired its last P2V in 2017, it donated remaining ones to museums in California, Oregon, Michigan, and Missoula.

idea where you were going to be when the day ended. I could wake up in one place and sleep three or four states away. I loved doing that and constantly moving around.

I enjoy doing this for Neptune. They'll bend over backward to help anyone on our team and demonstrated that when the daughter of one of our pilots burned herself. The company flew our corporate jet to rural eastern Montana to pick her up and fly her to a hospital burn unit. That's always been the mentality at Neptune, and it means a lot to everyone who works here. Today, as the maintenance control manager, I am responsible for everything on the maintenance side when the planes are deployed. I've been doing this the last thirteen years, but I still go out once a year to visit our crews.

Adrian Sanchez, Quality Assurance Inspector

Being part of aviation wasn't something that I planned for; it just kind of happened. A friend of mine was taking an AMP (Airframe and Powerplant) course, and I sat in on a class and decided to sign up the next semester. I was working in Washington when an acquaintance joined Neptune and then encouraged me to send in my resume, and Neptune brought me on to the team.

My first role was as a mechanic on the floor, starting in February 2018. We were doing heavy maintenance on aircraft during the off-season, and I enjoyed it. But then the next year, I was asked if I would be interested in becoming a crew chief. I started doing a little on-site training in May 2019, then went out on an airtanker in June. When you go out with a team now, it's for a month at a time, but in 2019 the total time we would be on the road was typically three to four months. We would work six days a week, and then our team would have a day off to recharge.

Serving as the crew chief is exciting work — it's also nerve-racking, but you get used to it. The key to being successful is having good communications with the crew and having captains who trust the crew chief and the people they work with. We hold each other's lives in our hands, and we all must have some level of trust to get the job done and to get home safely. I have to trust the pilots are doing their job because I can't fly the plane, and if I see a safety issue with the flight, they have to trust me and instinctively take action, and then ask questions later, since it's my team's job to make the plane safe to fly.

There were a lot of long days, but we made time to enjoy the downtime, visiting battlegrounds in Chattanooga, going fishing with the flight crew in Oregon, and we would make the time to hang out with each other and build that camaraderie.

When we come home, there is definitely an adjustment period. Your mindframe changes. For months it's you, a second mechanic, and the flight crew — a little family on the road, and then you return and there's a different dynamic in the shop. On the home front, it's like you were living a single life on the road, then you come back and see the family that you missed. So, that first week back, you are trying to settle back into life at home and at the shop.

I now work as the quality assurance inspector, validating the safety of all our aircraft, and I've earned my bachelor's degree in aviation maintenance management. I love it here in Missoula. People are friendlier, I like the atmosphere, and I have yet to meet a stranger here.

Chris Smith, Avionics Manager

As the avionics manager at Neptune, I work on all kinds of aircraft, but I enjoy working on the firefighting side a lot, because I know we are making an impact on the world, helping to put out wildfires. Everyone here understands the mission and what needs to be done. What I do, in simple terms, is work on all the things that tell the plane where to go, helping the plane talk to the ground. When there are wiring issues, my team fixes them. Today's planes are getting more complicated. Where planes used to be mostly mechanical, hydraulic, and pneumatic, current planes are more electronically operated. A lot of them can fly themselves for a long part of the flight, and my job is to make that safe.

Thinking about why I work at Neptune and in this business, I go back to this time when we were bidding for our next contract back in 2015. We needed to get our fleet ready to fly to Boise for a fly-in, and there were six or seven of us who volunteered to work nights to get the job done. What we thought would be a challenging task actually wasn't that bad — we finished with plenty of time to spare and it was because we all did more than what was expected. I was helping with things that didn't involve avionics, and the mechanics were doing the same.

Eric Komberec, Museum of Mountain Flying

In this area of the country, wildfires are a big thing — I'd see the big red planes with the big red tails on them and thought they were awesome. This is a really rewarding yet demanding career. A few years ago, we received a call for a fire in Anderson Hill. That's right here in Montana, and Anderson Hill is above my family's cabin, a property my father had purchased back in the 1970s. By the time I reached the fire, it had already started burning the cabin, and I was doing what I could to help save my dad's property. Meanwhile, my dad, who was still working as a contractor for the U.S. Forest Service, was on the ground using a bulldozer to create break lines. I still had a job to do, and there were other cabins and properties to try to save from the fire, so I kept flying. Much of the cabin, and the fifty years of timber that had grown around the property, were almost completely destroyed. Those trees won't grow back in my lifetime, but hopefully my daughters can enjoy them as they grow over the next fifty years.

Matt Dauenberwer, Maintenance Controller

The first time I served as a crew chief was back in 2011, and I stayed in that role through 2015. I was involved in a number of memorable missions, but the one that made me really appreciate my job and what we do here at Neptune is a wildfire that broke out south of Salt Lake City in 2012.

I was serving as the crew chief for Tanker 44. You think of the old philosophical question, "If a tree falls in the woods and nobody is around, does it make a sound?" That's kind of what it is like sometimes in aerial firefighting; you're dropping fire retardant, but you're not sure what impact you're having.

This particular drop didn't have any significant meaning for me until one of our other captains sent me a link to a video. I open the link, and there's a mother who is frantically loading her kids and any belongings she could fit into her car. She was very emotional, and she's focused on this fire that is coming down the hillside as she is saying goodbye to her home. She's finishing getting her kids into the car, then turns to the camera to say goodbye. Just then, Tanker 44 comes screaming through, dropping retardant — and I remembered that drop specifically — we hit her backyard fence, the kiddie pool. And right as we finish the drop, she starts crying, saying, "They just saved my house."

When you're part of the airtanker crew, you work a lot of hours, and you are away from your family a lot — but seeing reactions like that makes it all worth it.

Thinking back to one of the fun times we had, when I started out, our fleet was comprised of the old P2Vs. A great plane, but not as fast as our newer fleet of BAe 146 aircraft. When we upgraded, we all of a sudden had one of the fastest fleets in the industry. A competitor that didn't particularly like Neptune was serving as the lead plane and tried to beat us to a fire, forgetting how fast we were now. Our captain had to remind him we were loaded and ready to go, and the competitor tried to beat us there. This time, though, we had to wait for him to show up, and his copilot was relaying to us how their captain was cursing us out for getting there first.

Nicholas Lynn, Vice President of Operations

My father started working for Neptune in 1993 as a copilot and mechanic. He was based at our headquarters in Missoula, and in late July of 1994, he died in an airtanker crash. My father and his colleague Bob Kelly were piloting T-04, a P2V-7, and crashed northwest of Missoula, Montana, while fighting fire. I was fourteen at the time, and my father's death had a profound impact on me. I had always been interested in aviation because of my father's joy and passion for it. But after the accident, I became more interested in aviation and various facets of the industry. I think my father's accident instilled a sense of duty in me that carries on to this day. We have a responsibility to anyone who has perished in an accident to learn from it and ensure that it doesn't happen again. That is part of who I am.

I started my career here in the summer of 1995, between my freshman and sophomore years of high school. The owner offered me a job sweeping floors, cleaning parts, helping mechanics with whatever I could, basically learning the ropes. I started working full-time when I attended college. Classes would be scheduled in the morning, and then I was working into the evening. Later, I started deploying in the field, working as part of the crew to support airplanes, handling inspections, cylinder changes — whatever maintenance needed to be done. So, I've been at Neptune for thirty years, and you really have to have a passion for it and enjoy what you're doing to stay in one place for that long. I have too much fun here to go find something different.

It's always exciting chasing a plane around to see the country. Growing up here in Montana, a lot of friends left so that they could see more of the United States — but my job at Neptune allowed me to do that, and more. I've been from coast to coast, supporting fires in California, Texas, Minnesota, and Virginia. When I moved up to VP of operations, I started travelling globally, visiting Peru, Bulgaria, and Djibouti, to name a few, to inspect airplanes that we wanted to purchase.

The fire missions offer the most excitement. You're in a massive airplane, helping to save people on the ground. One particular mission comes to mind when I look back during my time working as part of the crew. We were fighting a fire close to home here in Missoula. We held our briefing that morning to chart out our plan of attack, and then we were orbiting over the fire when we received a call telling us that we were being diverted twenty miles north to the I-90 Complex fire. Someone had been driving a truck, and their chain came loose, sparking little fires along the highway. As we were flying, we saw all of these, and we came upon a little town called Alberton, where another fire had started. We dropped retardant to help save the town, and to this day, when we drive by the city's water tower, I can still point to the red retardant and remember the town we saved.

Whenever you're in an aircraft dropping retardant to prevent fire from spreading, it's exciting, but the Alberton event sticks with me the most. It was close to my hometown, and the drop was difficult due to the high winds and terrain. I was the crew chief in back keeping an eye on everything, and that was a very rewarding experience.

As a crew chief, I was on the road for months at a time. I made the decision to move on from that because I wanted to see my kids growing up, but there is a great sense of pride sacrificing your own safety to help others.

Randy Henrickson, Procurement Manager

When I found out I was hired by Neptune Aviation back in 1998, I was on top of the world. For the previous twenty years, I had been working for an agricultural and petroleum company. I started out as a pump jockey, working full-service pumps, and gradually moved up in the organization, becoming a store manager for one of our largest retail locations. Then there was a change in upper management and a change in business philosophies. That's when I started to look for a new opportunity.

At some point after that, I heard that there was a job opening with Neptune Aviation for a parts manager position, and twenty-six years later, I am still here. Thinking back on my time at the company, we faced a lot of challenges. One of the largest happened in May 2004, when the U.S. Forest Service cancelled our airtanker contracts. For the next two years, we didn't know what the future was going to hold. I was nervous. Both my wife and I were working here, and we had just refinanced our house, so if we had to leave, it would have been tough.

But our owner, Marta Timmons, assured all employees that they were going to be taken care of, and for those of us who stayed, that's exactly what happened. Marta's generosity and the care that she had for her team made it possible. In 2006, the Forest Service contracts were reinstated, and that was a very happy day for everyone here.

I have a lot of fun working here. One hot summer evening, after everyone went home, I received a call from one of our captains, Brad, who said he had to come back to the hangar because his plane was covered in retardant. I was thinking something had blown up at the tanker base while they were filling it and that the plane was covered with retardant on the outside. When he pulled in, I didn't see any retardant on the outside — but retardant was pouring out of the rear hatch. It looked like something out of the movie *The Blob*. Red was *EVERYWHERE*, and I asked Brad what the hell had happened. Turned out that it was likely an air bubble in the line, and during the filling process, with hundreds of gallons being pumped every minute, the line couldn't handle it, so it blew apart the pipe fittings inside and aft of the wing beam in Tanker 12 P2V. So, with fire hose in hand, Brad, his copilot, and I scrubbed out the rear of Tanker 12 and were absolutely covered in it. We were there for hours hosing out the plane and putting the manifold together again. Brad's an easygoing guy, but he was not pleased, especially having to wash his clothes about fourteen times before they were clean again. It wasn't funny when it was happening, but it was a great excuse to kick back and drink a beer to laugh about it afterward.

• • •

The overarching theme in Neptune Aviation is family — as it is in Norman Maclean's book. He wrote of the beauty of Montana and the complexities of families, logging, and fighting forest fires. His poignant quote, "We can love completely without complete understanding," might be the ethos of this Missoula aerial firefighting company.

Notes

Preface

1 Jeff Goodell, *Heat: Life and Death on a Scorched Planet* (New York: Black Inc., 2023).

2 National Interagency Fire Center.

3 Stephen J. Pyne, *Awful Splendour: A Fire History of Canada* (Vancouver: UBC Press, 2008).

4 Al Gore, "The Climate Crisis Is the Battle of Our Time, and We Can Win," *New York Times*, nytimes.com/2019/09/20/opinion/al-gore-climate-change.html.

5 Smokejumpers, firefighters who are inserted at the site of the fire by parachute, provide an initial response to remote wildfires. They are able to access remote fires in their early stages without needing to hike long distances carrying equipment and supplies. Smokejumpers are able to reach burn sites more quickly than land-based crews, and the aircraft can transport more equipment and supplies than could be carried to a fire by pedestrian crews.

Chapter 1: "Some Say the World Will End in Fire"

1 Robert Frost, "Fire and Ice," *New Hampshire*, 1923.

2 U.S. Forest Service, "Driving Through Washington's Largest Wildfire: A Self-Guided Tour of the Yacolt Burn," 2003, fs.usda.gov/Internet/FSE_DOCUMENTS/ fsbdev3_004902.pdf.

3 M.R. O'Connor, *Ignition: Lighting Fires in a Burning World* (New York: Bold Type Books, 2023), 7.

4 Buffalo soldiers were United States Army regiments composed exclusively of African American soldiers, formed after the Civil War to serve on the American frontier.

5 It's about time that Canada had an equivalent to Smokey Bear! In 2005, when a forest fire struck near his home, Burny the Firefighting Beaver became involved. No, he didn't suppress the fire by slapping it out with his tail, but as Nature's engineer, Burny quickly built a dam in a nearby stream to create a water source for the firefighters. With the fire extinguished, the fire chief thanked Burny for his hard work and offered him a job!

6 W.K. Fullerton, "National Forest Conferences in Historical Perspective," *Forestry Chronicle* 62, no. 5 (October 1986): 467–69. "The Convention can readily take credit for or was a factor in the establishment of professional schools at the Universities of Toronto, New Brunswick and Laval. The establishment of provincial forest services and fire suppression organizations was stimulated by this meeting."

7 Invented by the "Big Burn" hero "Big Ed" Pulaski, it was a sharp axe on one side and a mattock or grubbing blade on the other.

8 Peter J. Murphy, Cordy Tymstra, and Merle Massie, "The Great Fire of 1919: People and a Shared Firestorm in Alberta and Saskatchewan, Canada," *Forest History Today*, Spring/Fall 2015, 22–30.

Chapter 2: Getting a Bird's-Eye View: Early Attempts at Aerial Fire Detection

1 Forest History Society, foresthistory.org.

2 A historical marker was placed at that spot in 1955.

3 James Lewis, "June 29, 1915: First Aerial Fire Patrol Took Flight," Forest History Society, June 29, 2011, foresthistory.org/june-29-1915-first-aerial-fire-patrol-took-flight/.

4 Lewis, "June 29, 1915."

5 The British Columbia Aviation Museum at 1910 Norseman Road, North Saanich, V8L 5V5, has a full-scale replica of the Hoffar-1.

Chapter 3: Sentinels of the Forest Before Aviation

1 Justin Housman, "Hallie Daggett, First Woman Fire Lookout for USFS, Blew Minds: 'I Hope Your Heart Is Strong Enough to Stand the Shock,'" *Adventure Journal*, May 5, 2023, adventure-journal.com/2023.

2 The heliograph was a signalling device by which sunlight is reflected by mirror in flashes for messages in Morse Code.
3 Philip Connors, "A Talent for Sloth," Lapham's Quarterly, laphamsquarterly.org/lines-work/talent-sloth.
4 Lance Garland, "The Last of the Fire Watchers," Backpacker, May 16, 2023, backpacker.com/stories/last-fire-lookout-north-cascades-national-park/.
5 "Hallie M. Daggett: Early Woman Lookout," USDA Forest Service, fs.usda.gov/learn/our-history/women/hallie-m-daggett.
6 Jennifer McCartney, "The History of the Fire Tower: From World War II to Jack Kerouac to Temagami," Northern Ontario Travel, July 5, 2024, northernontario.travel/northeastern-ontario/history-fire-tower-temagami.

Chapter 4: The First Use of Aircraft to Detect Forest Fires

1 The author's *The Golden Age of Flying Boats* covers Curtiss's trials and the complete history of those aircraft.
2 The "H" was for Hammondsport, the original home of his company, the "S" for seaplane, and "L" for the Liberty engine.
3 Bruce West, "The Firebirds," Ontario Department of Lands and Forests, 1974, 2.
4 "Boston Expedition Finds Much Pulp Wood Accessible for Exploitation," *New York Times*, August 23, 1919, nytimes.com/1919/08/23/archives/airplanes-disclose-labradors-resources-boston-expedition-finds-much.html.
5 Designed in 1917 by Lieutenant Commander John Cyril Porte at the Felixstowe Naval Station in Britain and powered by two Rolls-Royce Eagle engines, the F.3 was an unwieldy colossus. Larger than the HS-2L, it could carry more but had less agility.
6 Hugh A. Halliday, "The Forest Watchers: Air Force, Part 35," *Legion Magazine*, October 20, 2009, legionmagazine.com/the-forest-watchers-air-force-part-35/.
7 John Parminter, "Introduction Aircraft and Their Use in Forestry in B.C.: 1918–1926," Academia.edu, August 1985, academia.edu/108582132/Introduction_AIRCRAFT_AND_THEIR_USE_IN_FORESTRY_IN_B_C_1918_1926_.
8 Two of those VIP passengers were Mrs. Jessie Meighen on November 5, 1920, and the next day, the Hon. Arthur Meighen, prime minister of Canada, both on demonstration flights from Kamloops. Arthur Meighen thus became the first Canadian prime minister to fly in a plane.
9 *Whistle Punk: B.C. Forest History Magazine* 1, no. 4 (Spring 1986).

10 Richard A. Rajala, "Feds, Forests, and Fire: A Century of Canadian Forestry Innovation," Government of Canada Publications, publications.gc.ca/collections/collection_2018/mstc-cstm/NM97-2-13-eng.pdf.

11 Considering each HS-2L had cost the U.S. taxpayer $30,000 (average cost per unit) when supplied to the Coast Guard, this was a bargain.

12 From the author's *Flying Canucks: Famous Canadian Aviators* (Toronto: Dundurn Press, 1994).

13 Stuart Graham, "The First Flying Patrol of Forests," *Canadian Forestry Journal* 15, no.5 (June 1919): 243–44.

14 In 2012, Parks Canada unveiled a commemorative plaque at Lac-à-la-Tortue (Turtle Lake). It reads: "The era of commercial bush flying began in 1919 at Lac-à-la-Tortue when Stuart Graham successfully piloted the first forest-fire reconnaissance flight in Canada, a feat that led to the creation of Laurentide Air Service Ltd."

15 Romeo Vachon returned to the site, having fallen in love with Georgette, one of the Tremblay girls, and they were married in October 1924. Canadian aviation owes much to both husband and wife.

16 R.W. Ayres, "The Summer of Twenty-Four," *Newhall Signal*, October 24, 1924, scvhistory.com/scvhistory/files/sg1924/sg19241024/sg19241024_orig.pdf.

17 In addition to the Vedette weathervane on the Confederation Building in Ottawa, there is also a plaque featuring a Vedette at the entrance to the Lac du Bonnet Airport, Lac du Bonnet, Manitoba, from where, in 1927, RCAF aircrews operated Vedettes.

18 After the war, many pilots kept their Royal Flying Corps ranks. It instilled a hierarchy and brought military discipline to the Air Service.

19 That Sault Ste. Marie was also the riding and residence of Lyons would return to haunt both men later.

20 The hotel's chef at the time was gameshow host Alex Trebek's father.

21 Halliday, "The Forest Watchers: Air Force, Part 35."

22 There are two rebuilt Vickers Vedettes, one at the Royal Aviation Museum of Westen Canada in Winnipeg and a second at the Western Development Museum in Saskatoon.

23 Restored as CF-ADH and on display at the Canadian Bush Plane Heritage Centre, Sault Ste. Marie, Ontario.

24 One of the Stinsons that the OPAS purchased was CF-OBA, which had been built for movie stars Clark Gable and Carole Lombard.

Chapter 5: Smokejumping

1 "The Fire of '33," *Glendale News-Press*, October 3, 1933, Los Angeles Fire Department Historical Archive, lafire.com/famous_fires/1933-1003_GriffithParkFire/1933-1003_GriffithParkFire.htm.

2 Frederick A. Johnsen, "Attacking Fires from the Air," *Air & Space Forces Magazine*, October 1, 2010, airandspaceforces.com/article/1010fire/.

3 My father fought with the Chindits in Burma (now Myanmar) during the Second World War and recalled that Brigadier Orde Wingate relied on mules to supply his own unit behind Japanese lines. Although the mules were hard-working, their braying gave positions away.

4 Like an aircraft, an autogiro is moved by an engine-powered propeller, but like a helicopter, lift is provided by a rotor. The rotor is not powered, and while the aircraft can land vertically, it cannot take off vertically.

5 An autogiro is supported in flight by unpowered rotating horizontal wings (or blades); forward propulsion is provided by a conventional propeller.

6 "History of Smokejumping," United States Department of Agriculture, Forest Service, Northern Region, Missoula, Montana: Aviation & Fire Management Northern Region, August 1, 1980, foresthistory.org/wp-content/uploads/2017/01/History-of-Smoke-Jumping.pdf.

7 P.D. Hanson (Regional Forester, U. S. Forest Service, Missoula, Montana), and Chas. L. Tebbe (Director, Northern Rocky Mountain Forest and Range Experiment Station, Missoula, Montana), *Aerial Bombing of Forest Fires: A Progress Report of an Experiment by the Army Air Forces and the U.S. Forest Services* (Missoula, MT: USDA Forest Service, Northern Region, August 1947), fs.usda.gov/rm/fire/wfcs/documents/further_reading/NWST-00776_Aerial%20Bombing%20of%20Forests_1947.pdf.

8 Smoke chasers are lightly equipped first responders who get to the fire early. The movie *Forest Smokechaser* explains the role. ffla-sandiego.org/fire-lookout/historic-topics/1948-forest-service-smoke-chasers/.

9 Not only is the Missoula Smokejumper Visitor Center here, but also the Missoula Smokejumper Base and the Missoula Technology and Development Center.

10 William D. Moody, *A History of the North Cascades Smokejumper Base*, Publications on Smokejumping, 4 (Missoula, MT: National Smokejumper Association, 2019), dc.ewu.edu/cgi/viewcontent.cgi?article=1003&context=smokejumping_pubs.

11 Fred Poyner IV, "North Cascades Smokejumper Base (Winthrop)," HistoryLink.org Essay 20747, March 26, 2019, historylink.org/File/20747.

12 Poyner, "North Cascades Smokejumper Base (Winthrop)."

13 Poyner, "North Cascades Smokejumper Base (Winthrop)."

14 Tom Decker, "Idaho City Jumps the Fires of Hell," *Smokejumper Quarterly Magazine*, October 2001, 18, smokejumpers.com/documents/magazine _pdfs/smokejumper-2001-10-3.1MB.pdf.

15 Gregg Phifer (Missoula '44), "My Brush with History/CPS 103 Jumpers," *National Smokejumper Association Quarterly Magazine*, October 2001, 3–4.

16 By 1944, there were so many German and Italian prisoners of war around that consideration was being given to use them as wildland firefighters. But the farmers and ranchers had first choice, and the Forest Service didn't get any.

17 Gregg Phifer (Missoula '44), "My Brush with History/CPS 103 Jumpers," *National Smokejumper Association Quarterly Magazine*, October 2001, 3–4.

18 The image of a buffalo adorned the reverse of the U.S. nickel from 1913 to 1938.

19 Scott C. Woodard, AMEDD Center of History and Heritage, "Medic Malvin L. Brown, First Casualty in the US Forest Service Smokejumper Program," February 23, 2016, army.mil/article/162507/medic_malvin_l _brown_first_casualty_in_the_us_forest_service_smokejumper_program.

20 Sherry Devlin, "Jumping into History," Missoulian, June 28, 1999, missoulian.com/jumping-into-history/article_8afb738e-0ff4-5b8b-ba4e-7e8edbbeb730.html.

21 Norman Maclean's "Young Men and Fire" tells the story of the Mann Gulch Fire. The author is better known for *A River Runs Through It and Other Stories,* a hauntingly beautiful book and movie, both favourites of mine. The book was set on the Missouri River, just one gulch downstream from the Mann Gulch fire.

22 For the complete paper, published June 1956, see D.W. Kelly, "The Saskatchewan Smokejumpers," The Forestry Chronicle, pubs.cif-ifc.org/doi /pdf/10.5558/tfc32230-2.

23 The Norseman CF-SAM that flew the smokejumpers is on exhibit at the Western Development Museum in Moose Jaw.

24 "The idea of actually parachuting into fires was a Soviet invention," says American wildfire historian Stephen J. Pyne. "In the 1930s these guys would climb out onto the wing of a plane, jump off, land in the nearest village, and rally the villagers to go fight the fire." Stephen J. Pyne, *Awful Splendour.*

25 For the same reasons, the military has not used its airborne divisions on a broad scale since the Second World War.

Chapter 6: Water Bombers: A Beginning

1 Bruce West, *The Firebirds* (Ontario Department of Lands and Forests, 1974), 215–31.

2 West, *The Firebirds*, 215.

3 West, *The Firebirds*, 217.

4 No date exists for this, but it must have been before the KR-34 was sold, in late 1944.

5 A.P. Heathcote, "Wings over the Wilderness: The Story of the OPAS," *Canadian Aviation Historical Society* Vol. 5, no. 1 (Spring 1967): 4, cahs.ca/component/virtuemart/1968-1963-back-issues/1967-volume-5/vol-5-no-1-spring-detail.

6 The thin metal tanks were war surplus material used by fighter planes to extend their fuel to cover longer range missions.

Chapter 7: Postwar Water Bombers

1 Fires that are suppressed in Canada are named. The most memorable one was the Kootenay Fire in 2003, which was called the "Holy Shit" fire because every time an official got up in a helicopter to have a look, the response was always the same: "Holy shit." In California, they had the "Charmin Fire" in 1997 when a female hiker burned her toilet paper instead of burying it. It burned eighteen thousand acres and cost $28 million to put out, making it the most expensive shit in history.

2 In later waterbombing reincarnations, it carried six hundred gallons, equally divided between two tanks that the pilot controlled by pull handles in the cockpit. The greatest user of Avengers as fire bombers in Canada was Forest Protection Limited (FPL) of Fredericton, New Brunswick. Buying twelve surplus TBMs from the Royal Canadian Navy in 1958, the company had them rebuilt as water bombers. By 1971, the FPL Avenger fleet had peaked at forty-three aircraft.

3 Emails from Tom Wilson.

4 One of the original Skyway Tiger Moths hangs from the ceiling of the Farm Machinery Museum in Langley.

5 Claire Palmer, "Historic Martin Mars Water Bomber Completes Final Flight in B.C.," CBC News, last updated August 12, 2024, cbc.ca/news/canada/british-columbia/hawaii-martin-mars-last-flight-1.7291472.

6 Daniel Erskine McIvor died on February 24, 2005, at the age of ninety-three. He was inducted as a member of Canada's Aviation Hall of Fame in 2002 and invested as a member of the Order of Canada in 2004.

Chapter 8: Birddogging

1 The air attack pilots I asked thought "Ride of the Valkyries" too clichéd but liked my choice, Creedance Clearwater Revival's "Bad Moon Rising." They thought Pink Floyd better, and if Canadian, "The Hip" (The Tragically Hip).

2 Seismic lines are narrow corridors cut out of the forest used to transport and deploy geophysical survey equipment.

Chapter 9: Catalinas and Cansos

1 In 1979, Knox Hawkshaw was awarded the J.D. McCurdy Award by the Canadian Aeronautical Institute for his contribution to Canadian aviation, the highest honour in the field of Canadian aeronautical engineering.

2 A porpoise (a bounced landing resulting in a series of jumps and dives) is caused when the seaplane's nose is either too high or too low because of incorrect pilot input on the elevator.

3 This occurred on May 21, 1978, when a Flying Fireman Canso CF-NTL crashed during a waterbombing mission when it lost control and nose-dived. The loss of control was caused by fuel starvation on the right engine.

4 William Shakespeare, *Henry V*, Act IV, Scene III.

Chapter 10: The Canadairs

1 In Spain, foresters believed that dropping salt water on their forests made the ground sterile. Spanish government agronomists had to persuade them that salt water was good for their forests.

2 In 2024, there were still fourteen Canadairs in Spain, all operated by the Air Force as part of Grupo 43. The CL-215Ts, which make up the bulk of the contingent with eleven units, are under the Ministry of Agriculture. They are complemented by three CL-415s under the Ministry of Defence.

3 Contrary to urban legend, the three-by-five-inch probes are much too small to ingest a scuba diver.

Chapter 11: Conair

1 "Our History," Conair Aerial Firefighting, conair.ca/about/our-history.

2 In 1992, Les Kerr was awarded one of Canada's highest aviation honours, the Trans-Canada (McKee) Trophy.

3 Conair at one time owned the former Canadian Pacific Airlines DC-6 that had brought the body of aviation icon Grant McConachie home to Vancouver from Seattle.

4 Ryan Mason, "The Last Mission," Aerial Fire, January 3, 2023, aerialfiremag.com/2023/01/03/the-last-mission/.

Chapter 12: Unsung Heroes

1 They did have FAA-issued-type certificates that specified they be maintained in accordance with any applicable military technical orders.

2 Blue Ribbon Panel Report to the Chief, USDA Forest Service and Director, USDI Bureau of Land Management, "Federal Aerial Firefighting: Assessing Safety and Effectiveness," Wildfire Today, wildfiretoday.com/documents/Blue_Ribbon_Panel_Final_12-05-2002.pdf.

3 The NTSB had recommended maintenance and inspection programs for firefighting aircraft, not cancelling the contracts.

Chapter 13: Helicopters and Air Tractors

1 Bob Petite, "Firestarters," Vertical, June/July 2009, helicopterheritagecanada.com/pdfs/V8I3-Rewind.pdf.

2 If the Bell 47 has come to symbolize the Korean War, then Bell's other legacy, the UH-1 Iroquois, the "Huey," transformed U.S. Army aviation during the Vietnam War, becoming one of the most recognizable aircraft in history.

3 Igor Sikorsky, AZ Quotes, azquotes.com/author/19661-Igor_Sikorsky#google_vignette.

4 Bill Gabbert, "1961 Higgins Ridge Fire — 20 Smokejumpers Were Rescued by a Tiny Helicopter," Wildfire Today, March 29, 2020, wildfiretoday.com/2020/03/29/1961-higgins-ridge-fire-20-smokejumpers-were-rescued-by-tiny-helicopter/.

5 Rodney D. "Rod" Snider was awarded the American Forest Service Medal for heroism and honored for saving the men's lives. In 2002, he was inducted into the Museum of Mountain Flying Hall of Fame.

6 In 1984, Terry Dixon was awarded the Star of Courage for conspicuous courage during the helicopter rescue of the survivors of a plane crash near Chilliwack, British Columbia.

7 This is a wildfire of such size, complexity, and/or priority that it requires a big team, heavy investment, and a long period of firefighting time to bring it under control.

8 Zac Delaney, "Helicopter Pilots Work to Get Jasper National Park Wildfires Under Control," *Edmonton Journal*, July 24, 2024, edmontonjournal.com/news/local-news/jasper-national-park-wildfires-helicopter-pilots.

9 In 2008, Leland Snow turned over his company to his employees. An avid runner and fitness enthusiast, he died on February 20, 2011, while jogging near his home.

10 The National Transportation Safety Board released the preliminary report into the crash that took her life. "The accident airplane was in the No. 2 position. During its first scoop sequence, witnesses on the lake and the pilots of the 2 SEATs flying behind the accident airplane saw the airplane make a left turn to the southwest. Subsequently the airplane impacted a vertical rockface bordering the southern shoreline of the lake, fell into the lake, and sank."

Epilogue: The Gods Didn't Warn Us

1 Lorna Crozier, "The Gods Don't Tell Us Everything," in "The Gods Didn't Warn Us": A Poem About the B.C. Wildfires, The Sunday Magazine, CBC, last modified July 16, 2017, cbc.ca/radio/sunday/july-16-2017-the-sunday-edition-1.4201674/the-gods-didn-t-warn-us-a-poem-about-the-b-c-wildfires-1.4201678.

2 Stephen J. Pyne, *Awful Splendour.*

Acknowledgements

1 Stephen King, *On Writing: A Memoir of the Craft* (New York: Scribner, 2020).

2 Christy Lefteri, *The Beekeeper of Aleppo* (New York: Ballantine Books, 2019).

Bibliography

Alexander, Linc W. *Fire Bomber into Hell: A Story of Survival in a Deadly Occupation*. Trenton, GA: Booklocker.com, 2010.

Asher, Damien, with Mouallem, Omar. *Inside the Inferno: A Firefighter's Story of the Brotherhood That Saved Fort McMurray*. Toronto: Simon & Schuster, 2017.

Blue Ribbon Panel Report to the Chief, USDA Forest Service and Director, USDI Bureau of Land Management. "Federal Aerial Firefighting: Assessing Safety and Effectiveness." Wildfire Today. wildfiretoday.com/documents/Blue_Ribbon_Panel_Final_12-05-2002.pdf.

Clark, S.A., A. Miller, and D.L. Hankins. "Good Fire: Current Barriers to the Expansion of Cultural Burning and Prescribed Fire in California and Recommended Solutions." 2021. karuktribeclimatechangeprojects.wordpress.com/wp-content/uploads/2021/03/karuk-prescribed-fire-rpt_final-1.pdf.

Connors, Philip. *Fire Season: Field Notes from a Wilderness Lookout*. New York: Harper Collins, 2012.

Egan, Timothy. *The Big Burn: Teddy Roosevelt and the Fire That Saved America*. New York: Houghton, Mifflin, Harcourt, 2009.

Gee, Alastair, and Dani Anguiano. *Fire in Paradise: An American Tragedy*. New York: W.W. Norton & Co., 2020.

Hansen, Heather. *Wildfire: On the Front Lines with Station 8*. Seattle: Mountaineers Books, 2018.

Lino, Amanda A. "How Forests and Forest Management Messaging Was Disseminated in Governmental Promotional Material in Ontario, 1800–1959." PhD diss., Lakehead University, Thunder Bay, Ontario, June 15, 2020.

Lustgarten, Abraham. *On the Move: The Overheating Earth and the Uprooting of America*. New York: Farrar, Straus & Giroux, 2024.

MacEachern, Alan. *The Miramichi Fire: A History*. Montreal and Kingston: McGill-Queen's University Press, 2020.

Macnutt, Alan. *Altimeter Rising: My 50 Years in the Cockpit*. Self-published, 2000.

Murphy, Peter. "History of Forest and Prairie Fire Control Policy in Alberta." Edmonton: Alberta Energy and Natural Resources, 1985.

Murphy, Peter J., Cordy Tymstra, and Merle Massie. "The Great Fire of 1919: People and a Shared Firestorm in Alberta and Saskatchewan, Canada." *Forest History Today*, Spring/Fall 2015.

Poyner, Fred, IV. "North Cascades Smokejumper Base (Winthrop)." Historylink.org Essay 20747, March 26, 2019. historylink.org/File/20747.

Pyne, Stephen J. *Awful Splendour: A Fire History of Canada*. Vancouver: UBC Press, 2008.

Pyne, Stephen J., and William Cronon. "The New World Order on Fire." In *World Fire: The Culture of Fire on Earth*, by Stephen J. Pyne. Seattle: University of Washington Press, 1997.

Roser, Max. "Technology Over the Long Run: Zoom Out to See How Dramatically the World Can Change Within a Lifetime." Ourworldindata.org, 2023. ourworldindata.org/technology-long-run.

Struzik, Edward. *Firestorm: How Wildfire Will Shape Our Future*. Washington, D.C.: Island Press, 2017.

United States Department of Agriculture, Forest Service, Northern Region. "History of Smokejumping." Missoula, Montana: Aviation & Fire Management Northern Region, August 1, 1980. foresthistory.org/wp-content/uploads/2017/01/History-of-Smoke-Jumping.pdf.

Vaillant, John. *Fire Weather: The Making of the Beast*. Toronto: Alfred A. Knopf, Canada, 2023.

West, Bruce. "The Firebirds." Toronto: Ontario Department of Lands and Forests, 1974.

Whistle Punk: B.C. Forest History Magazine 1, no. 4 (Spring 1986).

Willis, Clint, ed. *Fire Fighters: Stories of Survival from the Front Lines of Fire Fighting*. New York: Thunders Mouth Press, 2002.

Woodard, Scott C. "Medic Malvin L. Brown, First Casualty in the US Forest Service Smokejumper Program." AMEDD Center of History and Heritage, February 23, 2016.

Image Credits

Burton, Aaron, photographer: 103, 144
Caillet, Pierrick: xii, 119, 124, 126, 132
Cannon, Guy: 166
Conair: 134
Conair, photographer unknown: 151
Dixon, Terry: 157, 158
Forest Service, USDA: 56
Garrish, Tim: 109, 115
Glenn L. Martin Company: 91
Library & Archives Canada: 4, 16, 20, 30, 40
Neptune Aviation Services: 192, 196
Photographer unknown: 87, 88
Turchetti, Juliana: 175

Index

Locators in italics indicate images.

About the Author

Photo by Donna Hudson

A childhood at the end of a runway gave Canada's aviation author a lifelong sense of wonder of the miracle of flight itself. A career in Canada's Global Affairs led to a posting in Hong Kong, where watching the aircraft thread their way through the Kowloon tenements he wrote his first book, *Kai Tak: The History of Aviation in Hong Kong.* Returning home, he churned out in straightforward style the histories of Air Canada, Trans Canada Airlines, Canadian Airlines, and the engaging Flying Canucks series. In the tradition of author-pilots like Antoine de Saint Exupéry and Ernest K. Gann, whether writing about flying boats or feminism in aviation, Howard Hughes or Juan Trippe, Pigott has captured the possibilities and limits, the majesty and inequities of what he considers humankind's greatest technological achievement. *Fire Eaters* is his twenty-fifth book. Peter Pigott lives in Ottawa and is married to Donna Hudson.

Other Books by the Author

Air Canada: The History (2014)
American Obsession: Howard Hughes and Juan Trippe, Rivals in the Sky (2023)
Brace for Impact: Air Crashes and Aviation Safety (2016)
Canada in Afghanistan: The War So Far (2007)
Canada in Sudan: War Without Borders (2009)
Flying Canucks: Famous Canadian Aviators (1994)
Flying Canucks II: Pioneers of Canadian Aviation (1997)
Flying Canucks III: Famous Canadian Aviators (2000)
Flying Colours: A History of Commercial Aviation in Canada (1997)
From Far and Wide: A Complete History of Canada's Arctic Sovereignty (2011)
Gateways: Airports of Canada (1996)
Hong Kong Rising: The History of a Remarkable Place (1995)
Kai Tak: The History of Aviation in Hong Kong (1988)
National Treasure: The History of Trans Canada Airlines (2001)
On Canadian Wings: A Century of Flight (2004)
Royal Transport: An Inside Look at the History of British Royal Travel (2005)
Sailing Seven Seas: A History of the Canadian Pacific Line (2010)
See Jane Fly: Feminism in Aviation (2022)
Taming the Skies: A Celebration of Canadian Flight (2003)
The Golden Age of Flying Boats (2020)
Wings Across Canada: An Illustrated History of Canadian Aviation (2002)
Wingwalkers: The Rise and Fall of Canada's Other Airline (2003)
Wingwalkers: The Story of Canadian Airlines International (1998)